Studies in Applied Philosophy, Epistemology and Rational Ethics

Volume 34

About this Series

Studies in Applied Philosophy, Epistemology and Rational Ethics (SAPERE) publishes new developments and advances in all the fields of philosophy, epistemology, and ethics, bringing them together with a cluster of scientific disciplines and technological outcomes: from computer science to life sciences, from economics, law, and education to engineering, logic, and mathematics, from medicine to physics, human sciences, and politics. It aims at covering all the challenging philosophical and ethical themes of contemporary society, making them appropriately applicable to contemporary theoretical, methodological, and practical problems, impasses, controversies, and conflicts. The series includes monographs, lecture notes, selected contributions from specialized conferences and workshops as well as selected Ph.D. theses.

Advisory Board

More information about this series at http://www.springer.com/series/10087

Emiliano Ippoliti · Ping Chen
Editors

Methods and Finance

A Unifying View on Finance, Mathematics and Philosophy

Springer

Editors
Emiliano Ippoliti
Department of Philosophy
Sapienza University of Rome
Rome
Italy

Ping Chen
National School of Development
Peking University
Beijing
China

ISSN 2192-6255 ISSN 2192-6263 (electronic)
Studies in Applied Philosophy, Epistemology and Rational Ethics
ISBN 978-3-319-49871-3 ISBN 978-3-319-49872-0 (eBook)
DOI 10.1007/978-3-319-49872-0

Library of Congress Control Number: 2016957705

Printed on acid-free paper

This Springer imprint is published by Springer Nature
The registered company is Springer International Publishing AG
The registered company address is: Gewerbestrasse 11, 6330 Cham, Switzerland

Preface

Two views shape our understanding and approaches to finance, and stock markets in particular—as well as to a lot of other domains of human knowledge: the view from inside and the view from outside. This book examines these two views, analyzing their interconnections—when and where they clash, reconcile, or act independently—and the ways by which these views, in turn, affect various approaches to the study of finance and its branches. To this aim, the volume puts together contribution from experts from different fields: philosophy, economics, and physics. The core idea that motivates this choice is that the increasing complexity and pervasiveness of finance, and its computational turn, have to be dealt with several viewpoints.

These two views on finance employ radically different assumptions about the set of the relevant features of the domain and the way to look at them—e.g. the ontology, the methods, what counts as 'data', and the role of mathematical modeling. Accordingly, these views tackle very differently the two main sides of the problem of stock markets behavior—namely the quantitative and the qualitative side. The quantitative side requires figuring out an answer to prices' series—a mathematical tool to approximate them in the most cogent way. The qualitative side requires determining the variables that affect the dynamics of stock market prices, and their "machinery", as the result of the actions, or better interaction, of investors who sell and buy stocks.

But the view from inside and the view from outside are not necessarily conflictual. Mixed approaches emerge from a combination of the features of the two views. The two views serve goals and have scopes that might differ, but that can, and sometimes have to be used together since they can complement and extend each other. In effect, just like with triangulation, they offer distant viewpoints that can be combined in order to better map the land of financial phenomena. A conceptual triangulation that put in use a methodological pluralism.

The essays collected in this volume will provide new insights into, critical examinations of, and improvements to these approaches, dealing with both the quantitative and the qualitative approaches, the issue of prediction and the use of mathematics.

The volume is in two parts. The first looks at the view from outside, the second at the view from inside.

The first part of the volume starts with a description of the view from inside with the contribution Chapter "Methods and Finance: A View from Outside" by Emiliano Ippoliti. In his chapter Ippoliti emphasizes that that the outside view on finance maintains that that we can make sense of, and profit from stock markets' behavior, or at least few crucial properties of it, by crunching numbers and looking for patterns and regularities in certain sets of data. The basic idea is that there are general properties and behavior of stock markets that can be detected and studied through mathematical lens, and they do not depend so much on contextual or domain-specific factors. In this sense the financial systems can be studied and approached at different scales, and it is virtually impossible to produce *all* the equations describing at a *micro* level *all* the objects of the system and their relations. The typical view of the externalist approach is the one provided, for instance, by the application of statistical physics. In describing collective behavior, this discipline neglects all the conceptual and mathcmatical intricacies deriving from a detailed account of the inner, individual, and at micro level functioning of a system. The chapter examines how the view from outside deals with critical issues such as the mathematical modeling (Section "Mathematical Modeling"), the construction and interpretation of data (Section "Constructing and Interpreting Data"), the problem of prediction and performativity (Section "The Problem of Prediction and Performativity").

After this essay, five chapters highlighting important features of the view from outside follow.

In Chapter "Mathematical Representation in Economics and Finance: Philosophical Preference, Mathematical Simplicity, and Empirical Relevance" Ping Chen argues that, as pointed out by Keynes, classical economics is similar to Euclidean geometry. The problem is that the reality is non-Euclidean. Ping Chen argues that we have now robust evidence, and then we can take for granted, that market movements are nonlinear, non-equilibrium, and economic behavior is collective in nature. This contrasts the mainstream economics and econometrics, which are still dominated by linear, equilibrium models of representative agent. Moreover Chen argues that a critical issue in economics is the selection criteria among competing math models. He stresses that economists may choose the preferred math representation by philosophical preference, by mathematical beauty or by computational simplicity. He continues saying that he chooses the proper math of his models by its empirical relevance, even at the costs of increasing mathematical dimensionality and computational complexity. His main argument is that recent historical events of financial crisis reveal the comparative advantage of the choice of advanced math representations. Hi finishes arguing that technological progress eases future advancements in mathematical representation and philosophical change in economic thinking.

In Chapter "Behind the Price: On the Role of Agent's Reflexivity in Financial Market Microstructure" Fabrizio Lillo and Paolo Barucca review some recent results on the dynamics of price formation in financial markets and its relations to the efficient market hypothesis. Specifically, they present the limit order book mechanism for markets and they introduce the concepts of market impact and order flow, presenting their recently discovered empirical properties and discussing some possible interpretation in terms of agent's strategies. They argue that their analysis and findings support the idea that a quantitative analysis of data is crucial for the qualitative validation of hypothesis about investors' behavior in the regulated environment of order placement. They also argue that a quantitative analysis of data is crucial to connect micro-structural behaviors to the properties of the collective dynamics of the system as a whole, for instance market efficiency.

Finally they discuss the relation between some of the described properties and the theory of reflexivity. They basically propose that in the process of price formation both positive and negative feedback loops between the cognitive and manipulative function of agents are present.

In their Chapter "On the Description of Financial Markets: A Physicist's Viewpoint" Sergio Caprara and Francesco Sylos-Labini examine the main concepts of the neoclassical description of financial markets from a physicist's viewpoint. They start noting that, at least in the last two decades, physicists have devoted an increasing activity to the scrutiny of ideas and concepts that are at the basis of that branch of economic theory customarily called neoclassical. This activity, they continue, appeared as surprising in the early days of its rise and development, since the objects studied by physicists are usually atoms, molecules, planets or galaxies that look quite different from the objects studied by social sciences, the queen of which is economics. Human beings, they argue, on contrary to elementary particles or stars, are endowed with free will and, more importantly, the laws that rule the ways in which an individual makes her/his own choices and by which different individuals establish relations among them, developing a social behavior, are unknown to us. So, they conclude, it seems legitimate to doubt that such laws are well defined.

In their Chapter "Research Habits in Financial Modelling: The Case of Non-normality of Market Returns in the 1970s and the 1980s" Boudewijn de Bruin and Christian Walter consider finance at its very foundations, namely, at the place where assumptions are being made about the ways to measure the two key ingredients of finance: risk and return. It is well known, they emphasize, that returns for a large class of assets display a number of stylized facts that cannot be squared with the traditional views of 1960s financial economics (normality and continuity assumptions, i.e. Brownian representation of market dynamics). Despite the empirical counterevidence, they continue, normality and continuity assumptions were part and parcel of financial theory and practice, embedded in all financial practices and beliefs. Their gaol is to build on this puzzle for extracting some clues revealing the use of one research strategy in academic community, model tinkering defined as a particular research habit. We choose to focus on one specific moment of the scientific controversies in academic finance: the 'leptokurtic crisis', opened by Mandelbrot in 1962. The profoundness of the crisis, they note, come from the angle of the Mandelbrot's attack: not only he emphasized an empirical inadequacy of the Brownian representation, but also he argued for an inadequate grounding of this representation. They conclude their chapter by giving some insights into this crisis and displaying the model tinkering strategies of the financial academic community in the 1970s and the 1980s.

Matthias Leiss begins his Chapter "Super-Exponential Bubbles and Expectations: Theory, Simulations and Empirics" by reminding us that transient super-exponentially is a well-known statistical regularity of financial markets, which is generally associated with unsustainable growth and bubbles. His goal is to contribute to the understanding of super-exponential dynamics by assessing it from two new viewpoints. To this end, first he introduces an agent-based model of super-exponential bubbles on a risky asset market with fundamentalist and chartist traders. Then he shows analytically, and by simulations, that their mutual interactions repeatedly generate super-exponentially growing prices. Moreover, he combines our agent-based model with the methodology of log-periodic power law singularities (LPPLS) often used for bubble econometrics. Leiss introduces a new type of trader who employs the LPPLS framework to detect the existence of a bubble and invests accordingly, trying to ride the bubble while it lasts and to quit before the subsequent crash. His findings show that the presence of LPPLS traders increases market volatility. In the second part of the chapter, he construct risk-neutral return probability distributions from S&P 500 option prices over the decade 2003 to 2013 and argues that the data strongly suggest increasing option-implied return expectations prior to the crash of 2008, which translates into transient super-exponential growth expectations. Finally, he presents evidence for a regime-change from an abnormal pre-crisis period to a "new normal" post-crisis.

The second part of the volume, devoted to the view from inside, begins with an overall account of that view.

In his Chapter "Methods and Finance. A View from Inside" Emiliano Ippoliti characterizes this approach as the one maintaining that not only to study and understand, but also to profit from financial markets, it is necessary to get as much knowledge as possible about their internal structure and machinery. This view, Ippoliti stresses, argues that in order to solve the problems posed by finance, or at least a large part of them, we need first of all a qualitative analysis. In this sense the rules of trades executions, laws, institutions, regulators, the behavior and the psychology of traders and investors are the key elements to the understanding of finance, and stock markets in particular. Accordingly, data and their mathematical analysis are not the crucial element, since data are the output of a certain underlying structure of markets and their actors. The underlying structure is the ultimate object of the inquiry. Ippoliti in particular examines how the view from inside raises, and deals with, critical issues such as markets failure, information disclosure, and regulation; the notion of data, performativity, and the study of micro-structures.

After this introductory essay, four chapters highlighting important features of the view from inside follow.

In Chapter "The Sciences of Finance, Their Boundaries, Their Values" Alex Preda (King's College London) examines how several approaches from the social and natural sciences take finance, and especially financial markets as a domain of systematic inquiry. Preda notes that historians of economic thought have discussed extensively the emergence and evolution of some major, competing paradigms within finance, focusing on differences in their methodological and theoretical assumptions, as well as on the ways in which they have achieved dominant positions in the academia. Preda then consider a critical issue. More precisely: how these paradigms do see the problem of their own value, in relationship to the value of their field of study. In other words, he examines how they present finance as a set of phenomena worth studying, and what is valuable about studying them from a particular angle. To this end he examines five significant scientific approaches to finance: *financial economics*, *market microstructure*, *behavioral finance*, *social studies of science*, and *econophysics*. He shows how they represent the study of financial markets as a valuable, systematic endeavor, and how they represent their own value in providing a distinctive approach to the study of finance. He then distinguishes between internalistic and externalistic claims to value among these approaches. Internalistic value claims make reference to data accuracy and to methodological adequacy, while externalistic claims make reference to investigating links between finance and other forms of social organization and institutions.

In his Chapter "Quantification Machines and Artificial Agents in Global Finance: Historical-Phenomenological Perspectives from Philosophy and Sociology of Technology and Money" Mark Coeckelbergh raises questions about the societal, cultural and ethical significance of finance, mathematics, and financial-mathematical technologies, in particular the phenomenon of quantification as mediated by contemporary electronic information and communication technologies (ICTS). He first relates the history of mathematics to the history of financial technologies, and argues, inspired by Simmel and Marcuse, that from ancient times to now there seems to be an evolution towards increasing quantification not only in finance, accounting etc., but in modern society in general. His chapter then examines current shifts of financial agency that exemplify what seems to be a moment of hyper-quantification through the use of ICTs: experiences of "the market" as an independent agent and money machines as artificial agents in high frequency trading. The chapter ends by acknowledging the human character of finance and mathematics, warning that there are real human and social consequences of quantification, in ancient times and today, for society and responsibility.

In Chapter "Contemporary Finance as a Critical Cognitive Niche: An Epistemological Outlook on the Uncertain Effects of Contrasting Uncertainty" Lorenzo Magnani and Luca Bertolotti (University of Pavia) employ the cognitive niche construction theory, which aims at providing a new comprehensive account for the development of human cultural and social organization with respect to the management of their environment, to account for economic and financial phenomena. They start with their argument that cognitive niche construction can be seen as a way of lessening complexity and unpredictability of a given environment. In their chapter, they analyze economic systems as highly technological cognitive niches, and identify a link between cognitive niche construction, unpredictability and a specific kind of economic crises.

In Chapter "*Dark Data*. Some Methodological Issues in Finance" Emiliano Ippoliti argues that the nature of the data of financial systems raises several theoretical and methodological issues, which not only impact on finance, but have also philosophical and methodological implications, viz. on the very notion of data. In his chapter he examines several features of financial data, especially stock markets data: these features pose serious challenges to the interpretation and employment of stock markets data. In particular he focuses on two issues: (1) the way data are produced and shared, and (2) the way data are processed. The first raises an internal issue, while the second an external one. He argues that the process of construction and employment of the stock markets data exemplifies how data are theoretical objects and that "raw data" do not exist. Data are not clean, light and ready-to-use objects, and have to be handled very carefully and are a kind of "dark matter". *Dark data*, for the note.

Rome, Italy — Emiliano Ippoliti
Beijing, China — Ping Chen

Contents

Part I
The View from Outside

Methods and Finance: A View from Outside

Emiliano Ippoliti

Abstract The view from outside on finance maintains that we can make sense of, and profit from, stock markets' behavior, or at least few crucial properties of it, by crunching numbers and looking for patterns and regularities in certain sets of data. The basic idea is that there are general properties and behavior of stock markets that can be detected and studied through mathematical lens, and they do not depend so much on contextual or domain-specific factors. In this sense the financial systems can be studied and approached at different scales, since it is virtually impossible to produce *all* the equations describing at a *micro* level *all* the objects of the system and their relations. The typical view of the externalist approach is the one provided, for instance, by the application of statistical physics. By focusing on collective behaviors, statistical physics neglects all the conceptual and mathematical intricacies deriving from a detailed account of the inner, individual, and at micro level functions of a system. This chapter examines how the view from outside deals with critical issues such as the mathematical modeling (Sect. 2), the construction and interpretation of data (Sect. 3), and the problem of prediction and performativity (Sect. 4).

1 An Overview

Two views shape our understanding and approaches to finance, and stock markets in particular—as well as to a lot of other domains of human knowledge: the view from inside and the view from outside. They employ radically different assumptions about the set of the relevant features of the domain and the way to look at them—e.g. the ontology, the methods, and the role of mathematical modeling. The book examines these two views, analyzing their interconnections—when and where they clash, reconcile, or act independently—and the ways by which these views shape various approaches to the study of finance and its branches.

E. Ippoliti (✉)
Department of Philosophy, Sapienza University of Rome, Rome, Italy
e-mail: emi.ippoliti@gmail.com

E. Ippoliti and P. Chen (eds.), *Methods and Finance*,
Studies in Applied Philosophy, Epistemology and Rational Ethics 34,
DOI 10.1007/978-3-319-49872-0_1

Accordingly, these views tackle in a very different fashion the two main sides of the problem of stock markets prices behavior—namely the quantitative and the qualitative side. The quantitative side requires figuring out an answer to prices' series—that is a mathematical tool to approximate them in the most cogent way. The qualitative side requires determining the variables that affect the dynamics of stock market prices, and their "machinery", as the result of the actions, or better interaction, of investors who sell and buy stocks. The two sides, of course, are not totally separated and they interact in several ways on different problems.

In this chapter I will focus on the view from outside, or the externalist view.

The externalist view argues that we can make sense of, and profit from stock markets' behavior, or at least few crucial properties of it, by crunching numbers and looking for patterns and regularities in certain sets of data. The notion of data, hence, is a key element in such an understanding and the quantitative side of the problem is prominent—even if it does not mean that a qualitative analysis is ignored. As a matter of fact, few externalist approaches try to put together a qualitative analysis along with a quantitative one. But the point here that the outside view maintains that it provides a better understanding than the internalist view. To this end, it endorses a functional perspective on finance and stock markets in particular (see Bodie-Merton [3], p. 24–32). The focus on functions rather than entities or institutions seems to offer more stable objects of inquiry, from which the formers can be derived.

The basic idea of the externalist view is that there are general properties and behavior of stock markets that can be detected and studied through mathematical lens, and they do not depend so much on contextual or domain-specific factors. The point at stake here is that the financial systems can be studied and approached at different scales, and it is virtually impossible to produce *all* the equations describing at a *micro* level *all* the objects of the system and their relations. So, in response, this view focuses on those properties that allow us to get an understanding of the behavior of the systems at a *global* level without having to produce a detailed conceptual and mathematical account of the inner 'machinery' of the system. Here two roads open. The first one is to embrace an *emergentist* view (see e.g. Bedau—Humphreys [2, 21]) on stock market, that is a specific metaphysical, ontological, and methodological thesis: in this case it would be even useless to search for and write down all the equations for all the objects of the system and their relations at the micro level, since the properties at a different scale are not reducible to the one of the scales below it. The second one is to embrace a *heuristic* view, that is the idea that the choice to focus on those properties that are tractable by the mathematical models is a pure problem-solving option. This does not imply that we cannot account for the global behavior of the system in terms of its micro level dynamics and constituents, in the future. Maybe it requires a better understanding of the micro behavior of the systems, further data, and, in case, new pieces of mathematics.

A typical view of the externalist approach is the one provided, for instance, by statistical physics. In describing collective behavior, this discipline neglects all the conceptual and mathematical intricacies deriving from a detailed account of the inner, individual, and at micro level functioning of a system. Concepts such as stochastic dynamics, self-similarity, correlations (both short- and long-range), and scaling are tools to get this aim (see e.g. Mantegna and Stanley [12]).

Econophysics (see [6–9], [13, 15], [17, 18, 20], [21, 22]) is a stock example in this sense: it employs methods taken from mathematics and mathematical physics in order to detect and forecast the *driving forces* of stock markets and their *critical events*, such as bubbles, crashes and their tipping points [21, 22]. Under this respect, markets are not 'dark boxes': you can see their characteristics from the outside, or better you can see specific dynamics that shape the trends of stock markets deeply and for a long time. Moreover, these dynamics are complex in the technical sense, for they "can be understood by invoking the latest and most sophisticated concepts in modern science, that is, the theory of complex systems and of critical phenomena" ([21], xv). This means that this class of behavior is such to encompass timescales, ontology, types of agents, ecologies, regulations, laws, etc. and can be detected, even if not strictly predictable. We can focus on the stock markets as a whole, on few of their critical events, looking at the data of prices (or other indexes) and ignoring all the other details and factors since they will be absorbed in these global dynamics. So this view provides a look at stock markets such that not only they do not appear as a unintelligible casino where wild gamblers face each other, but that shows the reasons and the properties of a systems that serve mostly as a means of fluid transactions that enable and ease the functioning of free markets.

Moreover the study of complex systems theory and that of stock markets seem to offer mutual benefits. On one side, complex systems theory seems to offer a key to understand and break through some of the most salient stock markets' properties. On the other side, stock markets seem to provide a 'stress test' of the complexity theory. More precisely:

(1) stock markets are the quintessential of the so called *extreme events*, which highlight many social and natural system considered 'complex' in the technical sense;
(2) Stock markets, in particular their crashes, are paradigmatic of the emergence, in technical sense, of critical events in self-organizing systems.
(3) Stock markets' behavior challenge in a dramatic way our notion of forecast or prediction.

The analogies between stock markets and phase transitions, statistical mechanics, nonlinear dynamics, and disordered systems mold the view from outside, whose core is expressed by Didier Sornette:

> Take our personal life. We are not really interested in knowing in advance at what time we will go to a given store or drive to a highway. We are much more interested in forecasting the major bifurcations ahead of us, involving the few important things, like health, love, and work, that count for our happiness. Similarly, predicting the detailed evolution of complex systems has no real value, and the fact that we are taught that it is out of reach from a fundamental point of view does not exclude the more interesting possibility of predicting phases of evolutions of complex systems that really count, like the extreme events. It turns out that most complex systems in natural and social sciences do exhibit rare and sudden transitions that occur over time intervals that are short compared to the characteristic time

> scales of their posterior evolution. Such extreme events express more than anything else the underlying "forces" usually hidden by almost perfect balance and thus provide the potential for a better scientific understanding of complex systems [21], 17–18).

Phase transitions, critical points, extreme events seem to be so pervasive in stock markets that they are the crucial concepts to explain and, in case, foresee. And complexity theory provides us a fruitful reading key to understand their dynamics, namely their generation, growth and occurrence. Such a reading key proposes a clear-cut interpretation of them, which can be explained again by means of an analogy with physics, precisely with the *unstable* position of an object. A pencil held up vertically on your palm's hand is in a very unstable position that will lead to its fall sooner than later. It is just its position that leads to the fall, and not the instantaneous causes of the crash, such as a small motion of your hand, which is accidental.

Complexity theory suggests that critical or extreme events occurring at large scale are the outcome of interactions occurring at smaller scales. In the case of stock markets, this means that, unlike many approaches that attempt to account for crashes by searching for 'mechanisms' that work at very short time scales, complexity theory indicates that crashes have causes that date back months or year before it. This reading suggests that it is the increasing, *inner* interaction between the agents *inside* the markets that builds up the unstable dynamics (typically the financial bubbles) that eventually ends up with a critical event—the crash. But here the specific, final step that triggers the critical event—the collapse of the prices—is not the key for its understanding: a crash occurs because the markets are in an unstable phase and any small interference or event may trigger it. The bottom line: the trigger can be virtually *any* event *external* to the markets. The real cause of the crash is its overall unstable position—the proximate 'cause' is secondary and accidental. An advanced version of this approach (see [21, 22]) sees a crash as fundamentally endogenous in nature, whilst an exogenous, external, shock is simply the occasional triggering factors of it. The instability is built up by a cooperative behavior among traders, who imitate each other (in this sense is an endogenous process) and contribute to form and reinforce trends that converge up to a critical point. Additionally, in that case not only we have a dynamics that is detectable, but also exhibits a mathematical pattern that can be approximated by log-periodic functions (see Fig. 1)—or better a special class of them.

The main advantage of this approach is that the system (the market) would anticipate the crash by releasing precursory *fingerprints* observable in the stock market prices: the market prices contain information on impending crashes and this implies that:

> if the traders were to learn how to decipher and use this information, they would act on it and on the knowledge that others act on it; nevertheless, the crashes would still probably happen. Our results suggest a weaker form of the "weak efficient market hypothesis", according to which the market prices contain, in addition to the information generally available to all, subtle information formed by the global market that most or all individual traders have not yet learned to decipher and use. Instead of the usual interpretation of the efficient market hypothesis in which traders extract and consciously incorporate (by their

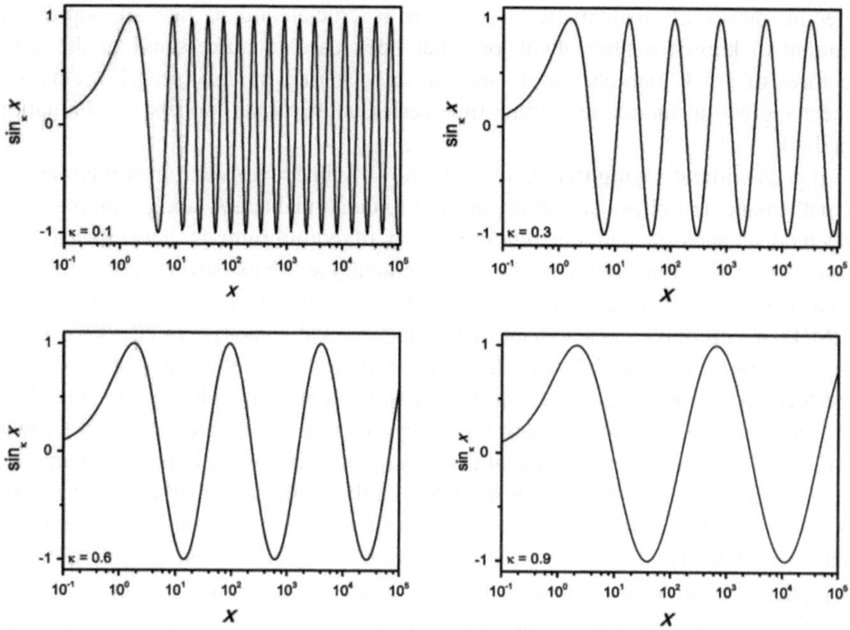

Fig. 1 Examples of log-periodic functions

> action) all information contained in the market prices, we propose that the market as a whole can exhibit "emergent" behavior not shared by any of its constituents [21], 279).

In a nutshell, the critical events emerge in a self-organized and cooperative fashion as the *macro* result of the internal and *micro* interactions of the traders—their imitation and mirroring. Thus, the employment of the notion of emergence has a strong heuristic power: if this hypothesis is right, we do not need to look at the many internal dynamics of system in order to understand its behavior, but only at few of its external features.

2 Mathematical Modeling

Mathematical modeling[1] plays a pivotal role in the view from outside, not only because it aims at answering the quantitative side of the problem of stock market prices, but especially because it maintains that it is *just* through the quantitative side of the problem that we can eventually find a solution to the general problem. By solution here we mean a way to foresee future trends in stock markets, or at least precursory signs of critical events—i.e. crashes. Accordingly, the choice of the

[1]See [16] for a detailed account for the role of models in economics.

specific pieces of mathematics to employ is crucial and in fact an impressive amount of known mathematical tools has been used to make sense of the main features of stock markets—tools such as agent-based models (see [1]), classical linear regression model, univariate time series, multivariate models, or simulation methods.

It goes without saying that this is not a novelty in economics: many theories use a mathematical or physical analogy as their foundational act. A stock example is the mechanical analogy and the relative notion of equilibrium in neoclassical economics (see MacLure [14]). Some of these analogies shape several approaches to finance as well. Let us think at the strong version of the Efficient Market Hypothesis (EMH), which draws on a variant of the 'mechanical analogy'. In effect the EMH employs the controversial hypothesis of a representative agent endowed with 'perfect rationality' and the utility maximization principle as a guide for decision-making. One of the main advantages of this 'ideal' agent, as well known, is mathematical in nature: this agent is perfectly treatable not only in a mathematical fashion but, above all, with linear models. But the choice of a mathematical model, as also Ping Chen points out in his contribution to this volume, in not a neutral act, but a 'philosophical' choice. It embeds, and accordingly can reveal, a lot of hypotheses made to conceptualize features of the stock markets. Using power-law distributions, correlations, scaling, and random processes as concepts to approach stock markets and their data, just to mention few examples, is very different from employing normal distributions, or random walks. Since during the past 30 years physicists have achieved important results in the field of phase transitions, statistical mechanics, nonlinear dynamics, and disordered systems, the application of this quantitative tools to financial systems is increasing a lot (see Mantegna and Stanley [12]).

These mathematical models, using a popular metaphor (see [19]), are means for searching a 'signal in the noise' of the data, that is to find an equation (a pattern) capable to fit the data in a cogent way. The problem here is the well-known underdetermination of equations by data: the same set of data can be approximated by several (infinite) equations. Since the equations provide only an approximation of the set of data, the choice of a specific equation over another one is not a pure logical or probabilistic act, but involves pragmatic criteria and a cost-benefits analysis. You loose something and you win something with each of these equations and the choice depends a lot on your specific purposes. In effect mathematical modeling serves several purposes in finance. In particular statistical techniques, especially econometrics, have played a great role in the mathematical modeling of stock markets' behavior. Financial econometrics (see in particular [4]) is employed for testing hypotheses or relationships between variables, for determining asset prices or returns, evaluating the impact on financial markets of changes in economic conditions, foreseeing plausible values of specific financial variables and for financial decision-making. A strong example of a mathematical model of this kind, is the one encapsulated in a log periodic equation, which is employed also to advance forecasts.[2] This model aims at predicting a critical event *c*—just like a crash. The hypothesis employed here is that since *c* is usually preceded by oscillations whose frequency increases as we approach the time t_c of its occurrence (following a Log-Periodic Power Law, LPPL), it is possible to determine t_c reasonably well by monitoring such oscillations.

3 Constructing and Interpreting Data

On one hand, the mathematical tools usually employed in financial analysis are basically the same as those used in economic analysis, on the other hand financial data, more often than not, are different from the macroeconomic data (w.r.t. frequency, seasonality, accuracy, etc.). While in economics we have problems such as the *small samples problem*, *measurement error* or *data revisions*, in finance they are much, much less common, if any.

In the first case, the *small samples problem*, for instance economic findings for government budget deficits, data are measured only on an annual basis, and if the technique used to measure these variables changed, let us say, a 25 year ago, then only twenty-five of these annual observations would be useful (at most). In the second and third case, the data might be estimated or measured with error, leading to subsequent revisions. In general these problems are rare in finance. In effect financial data are so that:

- prices and other variables are recorded as a transaction takes place, or when are quoted on the displays of information providers.
- some sets of financial data are recorded at much higher frequencies than macroeconomic data—at daily, hourly, and now even milliseconds-by-milliseconds frequencies. Here we enter the realm of 'big data' and the number of findings available for analysis can potentially be very large. This means that powerful techniques can more easily be applied to financial rather than to economic data, and the results are more reliable.

Moreover, financial data are not regular (in time): the price of common stocks for a company, for instance, generally is recorded whenever there is a new trade or quotation placed by the financial information recorder. Such recordings are very unlikely to be evenly distributed over time.[3] A standard way of dealing with this issue is to take a certain frequency and use the *last* prevailing price during the interval of it. Moreover a standard way of treating financial data is to employ all

[2]The equation, developed by Sornette, can be expressed in the following simple way: y(t) = A + B (tc − t) z + C (tc − t) z cos (ω log (tc − t) + Φ).

[3]Usually there is no activity when the market is closed, there is less activity around the opening and closing of the market, and around lunch time.

data of the same frequency. This means that if we are trying to estimate, for instance, a model for stock prices using weekly observations on macroeconomic variables, then we must also use weekly observations on stock returns, *even if* observations on the latter variable are recorded monthly.

Furthermore, the financial data exhibits other problems, as their 'noisy'—so that it is more complicated to draws apart underlying trends or patterns on one hand and random or uninteresting features on the other. Furthermore, the set of high frequency data often presents additional properties and dynamics, which are a consequence of the way the market works, or the way the prices are recorded.

In sum, the understanding of the notion of data, and their construction, in financial systems and stock markets in particular, is essential. Just to give an example of the complexity of financial data let us consider a much-talked piece of financial data, that is the alleged dimension of financial markets. How big are financial markets? A common figure, calculated on data provided by the Fed and World Bank, says us that the financial markets are very large, maybe too large: they are much larger than the economy—a multiple of it! For instance we know that in 2011:

(1) the total amount of debt in US was 54,2 trillions dollars (loans only—source: FED)
(2) the US stock market was 15,6 trillions dollars (source: World Bank)
(3) the US money supply was 10 trillions dollars (source: FED)

So, adding (1) and (2) we get a total about 70 trillions dollars as a good estimation of the dimension of financial markets. Moreover we know that:

(4) the GDP (Gross Domestic Product), that is the total value of the financial goods and services produced in an economy in one year (a benchmark for measuring the economy of a country), was about 15 trillions in 2011 for US.

So we have that US stock market (2) alone is more ore less equal to Us GDP (4)! And also that a great portion of US financial markets (1 + 2) is five times US GDP (4)!

Why do the stock markets seem so large? The answer is in the very construction of the data. In fact, here we are merging in a single piece of data (a number) two variables of different kind, namely **stock** (a) and **flow** (b) variables:

(a) the size of financial markets, an index of wealth, which measure the *current, static* value of *all* assets we own now, how much we have of something at a point in time.
(b) GDP, an index of income, which measures an *activity*, the money we earn each month/year, and hence how much we did something over a specific period of time.

The point here is that variable (a) has the problem of so-called *double* or *multiple counting*, that is the fact that a transaction is counted more than once. Money travels trough the financial markets: an income of one institutional unit is the expenditure of another, and what is an input of one institutional unit can be the output of another.

This passage, which often can be more than one, affects our data on the size of the financial markets. Let us consider for instance a typical situation: a household making a deposit in a local bank, say 10.000 €. This seemingly simple act can generate, and usually does, several assets. The local bank can make a loan to a regional bank, and this, in turn, lends money to a corporation that undertakes a big project. In this case the household deposit has passed along twice, i.e. three business have used these money to generate assets: each bank has a new asset (a loan), and the corporation has a new asset (a new project). In that case also new debts are generated: the one that the local bank owns to the depositor, the one the regional bank owns to the local bank, and the one the corporation owns to the regional bank. So the same amount of money, 10.000 €, generates a lot of multiple counting, that are recorded in the statistics and our data about financial markets. That explains why they seem so big: the more complex the path that money takes to get from the original lender to the ultimate borrower, the bigger the financial system will appear to be. Thus not only it is no surprising that financial markets are bigger then the economy, but in a sense they should be so—and this can also be good. They are really big and this means that the financial system (1) has backups, and (2) the there is more competition between the lenders, a fact that lowers the costs and improves the quality of the financial services.

Before pointing at a crucial property of stock markets data in particular, it is worth recalling here that there are basically three kinds of data in quantitative analysis of financial phenomena: *time series data*, *cross-sectional data*, and *panel data*.

Time series data are collected over a period of time for one or more variables. More precisely time series data are built over a particular *frequency of observation* —or data points. The frequency is simply a measure of the interval over which the data are recorded. For instances data about industrial production are recorded monthly or quarterly, while government budget deficit on annual base. The data may be quantitative (e.g. prices or number of shares outstanding), or qualitative (e.g. the day of the week or a credit rating, etc.). These data are employed to mine correlations where *time* is the most important dimension. For instance, they can be used to examine how the value of a stock index of a country changes as the macroeconomic fundamentals of it change, or how the value of a company's stock price has changed after the announcement of the value of its dividend payment.

At the other side of the temporal scale we find the *cross-sectional data*, that is data on one or more variables collected at a *single point in time* (hence without differences in time). These data offer an instant picture of a system and tell nothing about the history or the changes of it over time—for example, the data might be a sample of bond credit ratings for USA banks.

This kind of data are employed to tackle problems like the relationship between company size and the return to investing in its shares, the relationship between a country's GDP level or the probability that the government will default on its sovereign debt.

Then we find the *panel data*, which have the dimensions of both time series and cross-sections, e.g. the daily prices of a number of blue chip stocks over two years. Thus panel data provide observations of multiple phenomena over *multiple time periods* for the *same* entities (e.g. firms or individual). In this sense, time series and cross-sectional data can be seen as panel data of just one dimension.

We are now in a position to examine a crucial property of the data about stock markets. In effect these data come in a very peculiar form: we do not know *who* trades what. We have records for time, volume, price of stocks, but we miss this fundamental piece of information: trading-account identifiers are nearly always cut out and this makes almost impossible to detect sequences of actions by the same executing trader. This could be a serious obstacle in the understanding and modeling of stock markets and financial systems. How can we read the dynamics of a social system if we do non know who is doing what?

Here the view from outside seems to score a point. As a matter of fact, the absence of this kind of data is not a great concern for an externalist view: if this approach is right, you can ignore this piece of data to a large extent without compromising the understanding of crucial proprieties and dynamics of stock markets. Since it is the collective behavior that counts and will determine the overall dynamics of the system, the specific information of the individual behavior and choices can be safely ignored. What is possible at the individual level it is not possible at an aggregate level and vice versa. The peculiarities and possible deviances of a single individual's behavior are cancelled out and "absorbed" in the collective behavior.

It is not coincidence that a paradigmatic externalist approach, like the one put forward by Sornette, draws on an analogy with geology, and seismology in particular. Just as in seismology, where most events occur at a depth such that there is no way to directly measure the force (the stress), in stock markets we do not have a way (the data) of seeing the force acting in depth and at an individual level. We have to measure it indirectly, statistically, and collectively.

The point here is to what extent and under what conditions this piece of information can be considered as negligible for the understanding of stocks markets. The view from outside takes a clear-cut position, but the problem still remains open, at least in part. Maybe you can ignore this piece of data in the study of *certain* few dynamics of the stock markets, but in other cases you should consider this very carefully, especially in case of asymmetries, for instance when few actors are capable to affect the system or large part of it.

4 The Problem of Prediction and Performativity

The outside view goes under a stress test with the notion of prediction, which becomes so controversial in stock markets. One of the main theoretical problems in finance, and in the social sciences in general, as argued by Flanagan [10] and then by practitioners like George Soros [23], is its 'reflexivity': the fact that a prediction, and hence an expectation, about the behavior of the system is able to affect and change it. This is arguably one of the greatest differences between social sciences, where systems are able to learn and are reflexive, and natural sciences, where systems do not exhibit such a feature. This fact makes it very difficult to evaluate a financial hypothesis by means of a comparison between a prediction about it and the events that happen *in reality*, as the former can affect the latter. As an example, consider a prediction released into a financial system (e.g. the forecast that a crash of about 15–25% will occur in six weeks). It will *generate* one of the following three scenarios depending on the threshold of how many investors believe the prediction:

(a) *not enough* investors believe it, which then is useless, and the market drops just as predicted. Even if this seems a victory for the hypothesis that generates the prediction, criticism can define it as 'lucky one', which does not have any statistical significance.
(b) *A sufficient* number of investors believe it, who adjust their strategies accordingly and the bubble vanishes: the crash does not occur and the forecast is self-refuting.
(c) *A lot* of investors believe it, causing panic, and the market drops as a consequence. So the forecast is self-fulfilling and its success is due to the effect of the panic rather than to the predictive power of the hypothesis.

These scenarios are problematic for the theory that forecast them: in case (a) and (c) the crash is not avoided, while in (b) the forecast is self-refuting and accordingly the theory turns out to be unreliable. This feature seems to be a defining property of the learning and reflexive systems and raises also the problem of scientific responsibility, in particular the responsibility to publish scientific findings.[4]

This feature of the stock markets has been re-conceptualized and refined by Callon [5] and MacKenzie [11], who examined the notion of *performativity* in stock markets. In particular MacKenzie examined the ways in which the financial models and theories influence and shape the systems they seek to understand. These models, quoting Mackenzie, are an engine and not a camera of financial markets: "financial economics, I argue, did more than analyze markets; it altered them. It was an 'engine' in a sense not intended by Friedman: an active force transforming its environment, not a camera passively recording it" (MacKenzie [11], 12).

Now, performativity can be broken down in two parts. On one hand we find the so-called Barnesian performativity. It describes those situations where the practical employment of an economic theory makes economic phenomena resemble what they are described to be by that economic theory. On the other hand we find the so-called counter-performativity, where the employment of an economic model makes economic phenomena differ from the description of them given by this model. Nowadays, quantitative trading (QT) is done by most of the major players in the financial markets, especially the ones committed to an externalist view. QT of

[4]This issue becomes much more complex when researchers have to take into account the potential effect of the publication of their in society. The stock markets exemplify this problem and its complexity.

course is dependent on mathematical modeling, some of which implement sophisticated strategies. It follows that if the performativity hypothesis is valid, then those model are shaping the financial system—in both ways. The problem is to detect when a model or a theory is "performing" or not ("counter-performing"). For instance Alex Preda in his contribution to this volume underlines that Behavioral finance (BF) could have a counter-performative effect, since its employment would modify the behavior of its practitioners, making them resemble less its predictions: "paradoxically then, while financial economics operates within the ideal of emotionless, fully attentive, perfectly calibrated market actors, it would be behavioral finance which brings real market actors closer to this approach".

Of course, all this complicates a lot the nature, the role and the use of predictions in financial systems, in particular stock markets. If a prediction based on a certain model can be performative or counter-performative, it is hard to see when a model and its predictions are working properly.

A stock example of an externalist answer to this issue is the one put forward by econophysics, which seems to support a weak (weaker) version of the efficient market hypothesis (EMH) and draws on the hypothesis of existence of log-periodic structures in stock markets as the outcome of the existence of cooperative behavior among traders imitating each other. Such an approach 'saves' predictions and models in stock markets in the following way: "the market would anticipate the crash in a subtle self-organized and cooperative fashion, releasing precursory 'fingerprints' observable in the stock market prices" ([21], 279) and "even if the traders were to learn how to decipher and use this information [...] the crashes would still probably happen" (Ibid).

In a nutshell, such an answer argues that the market prices contain information on brewing critical events (crashes), and that this piece of information is formed by the global market as a whole. So "instead of the usual interpretation of the efficient market hypothesis in which traders extract and consciously incorporate (by their action) all information contained in the market prices, we propose that the market as a whole can exhibit "emergent" behavior not shared by any of its constituents" (*Ibid.*).

References

1. Abergel, F., Aoyama, H., Chakrabarti, B.K., Chakraborti, A., Ghosh, A.: Econophysics of Agent-Based Models, London: Springer (2014)
2. Bedau, M., Humphreys, P. (eds.): Emergence: Contemporary Readings in Philosophy and Science. Cambridge (Massachusetts): The MIT Press (2008)
3. Bodie, Z., Merton, R. Finance. Prentice-Hall, New Jersey (2000)
4. Brooks, C.: Introductory Econometrics for Finance. Cambridge University Press, New York (2008)
5. Callon, M.: Introduction: the Embeddedness of Economic Markets in Economics. In: Callon, M. (ed.) The Laws of the Markets. Blackwell, Oxford (1998)
6. Chatterjee, A., Chakraborti, A., Chakrabarti, B.K.: Econophysics and Sociophysics: Trends and Perspectives. Wiley-VCH, Berlin (2006)

7. Chen, P.: Equilibrium illusion, economic complexity, and evolutionary foundation of economic analysis. Evol. Inst. Econ. Rev. **5**(1), 81–127 (2008)
8. Chen, P.: Economic Complexity and Equilibrium Illusion: Essays on Market Instability and Macro Vitality. Routledge London (2010)
9. Chen, P.: Nonlinear Dynamics and Evolutionary Economics, (co-edited with Richard Day). Oxford University Press, Oxford (1993)
10. Flanagan, O.J.: Psychology, progress, and the problem of reflexivity: a study in the epistemological foundations of psychology. J. Hist. Behav. Sci. **17**, 375–386 (1981)
11. MacKenzie, D.: An Engine, Not a Camera How Financial Models Shape Markets. MIT Press, Boston (2006)
12. Mantegna, R., Stanley, H.: An Introduction to Econophysics: Correlations and Complexity in Finance, Cambrdige: Cambridge University Press Cambridge (1999)
13. McCauley, J.: Dynamics of Markets: Econophysics and Finance. Cambridge University Press, Cambridge (UK) (2004)
14. McLure, M.: Pareto, Economics and Society: The Mechanical Analogy. London: Routledge (2001)
15. Mirowski, P.: More Heat than Light—Economics as Social Physics, Physics as Nature's Economics. Cambridge University Press, Cambridge (UK) (1989)
16. Morgan, M.S.: The World in the Model. How Economists Work and Think. CUP, Cambridge (2012)
17. Ray, R.: Econophysics. Finance, economics and physics. Appl. Econ. Lett. **18**, 273–277 (2011)
18. Savoiu, G.: Econophysics: Background and Applications in Economics, Finance, and Sociophysics (2012)
19. Silver, N.: The Signal and the Noise. Penguin Books, New York (2012)
20. Sinha, S., Chatterjee, A., Chakraborti, A., Chakrabarti, B.K.: Econophysics: An Introduction. Wiley-VCH (2010)
21. Sornette, D.: Why Stock Markets Crash. Princeton University Press, Princeton (2003)
22. Sornette, D.: Physics and Financial Economics (1776–2013) Puzzles, Ising model and generalizations. RPP (2013)
23. Soros, G.: The Alchemy of Finance. Simon & Schuster, London (1988)

Mathematical Representation in Economics and Finance: Philosophical Preference, Mathematical Simplicity, and Empirical Relevance

Ping Chen

Abstract As Keynes pointed out, classical economics was similar to Euclidean geometry, but the reality is non-Euclidean. Now we have abundant evidence that market movements are nonlinear, non-equilibrium, and economic behavior is collective in nature. But mainstream economics and econometrics are still dominated by linear, equilibrium models of representative agent. A critical issue in economics is the selection criteria among competing math models. Economists may choose the preferred math representation by philosophical preference; or by mathematical beauty or computational simplicity. From historical lessons in physics, we choose the proper math by its empirical relevance, even at the costs of increasing mathematical dimensionality and computational complexity. Math representations can be judged by empirical features and historical implications. Recent historical events of financial crisis reveal the comparative advantage of the advanced math representation. Technology progress facilitates future advancements in mathematical representation and philosophical change in economic thinking.

1 Introduction: What's Wrong with Economic Math?

There is a quiet revolution in scientific paradigm that began in 1970s [1] that has increasing impact to mathematical economics since 1980s [2]. Advances in nonlinear dynamics and non-equilibrium physics fundamentally change our views from the origin of life to the evolution of universe. However, mainstream economics is reluctant to adopt new complexity science and its application to economics. Before and after the 2008 financial crisis, there is a strong criticism of excess mathematical economics and its effort in imitating physics [3–5]. However, few pointed out what's

For the Workshop on Finance, Mathematics, and Philosophy, Rome, June 12–13, 2014.

P. Chen (✉)
China Institute, Fudan University, Shanghai, China
e-mail: pchen@nsd.pku.edu.cn

E. Ippoliti and P. Chen (eds.), *Methods and Finance*,
Studies in Applied Philosophy, Epistemology and Rational Ethics 34,
DOI 10.1007/978-3-319-49872-0_2

wrong with economic math. Krugman blamed economics for "mistaking beauty for truth," [6] However; he did not elaborate how to judge beauty and truth in economics.

As trained as a physicist, a more fundamental question is: Why does economics need math? There are two possible answers: First, using math as a tool for discovering regularities from large numbers of numerical data, just like physicists or physicians do when analyzing experimental data. Second, using math as a language to express a concept or belief, so that people may accept economics as a branch of science in the modern era. This motivation is visible in neoclassical economics.

And then, we have another question: how to select a preferred model among several competing representations based on the same data set? We have two possible choices: we may select a math model based on its empirical relevance or mathematical beauty. This is the main issue discussed in this article.

Introducing math into economics is a development process. Before the Great Depression in 1930s, economic debate was mainly contested by historical stories and philosophical arguments. After the Great Depression, an increasing number of economic data has been collected by governments and corporations. Rising demand in data analyses stimulates rapid application of statistics and math in economic studies. The IS-LM model in neoclassical economics became a professional language in policy debate on fiscal and monetary policy. Furthermore, economic math began to dominate academia and universities since 1950s. The equilibrium school succeeded in promoting their belief in self-stabilizing market mainly by intellectual power in economic math. Their weapon in math representation is the unique equilibrium in linear demand and supply curves. In contrast, the Austrian school emphasized the organic nature of economic systems. They used philosophical arguments against neoclassical models. Since 1980s, the evolutionary school has been integrating nonlinear dynamics with evolutionary mechanisms.

Now we face a methodological and philosophical challenge: which school is better at understanding contemporary issues such as business cycles, economic crisis, economic growth, and stabilizing policy. And which mathematical representation is a better tool in quantitative research in economics and finance.

In this article, we will demonstrate different math representations and competing economic thoughts, and discuss related issues in economic philosophy.

2 Mathematical Representations of Competing Economic Thoughts

The central issue in economics and finance is the cause of market fluctuations and government policy in dealing with business cycles. The equilibrium school believes that market movements are self-stabilizing; therefore the laissez-fare policy is the best option. The disequilibrium school may use fiscal and monetary policy when market is out of equilibrium in a short-term. The non-equilibrium school studies more complex situations including economic chaos and multiple regimes. Math models are widely used in economic debate since 1930s.

Competing economic math is closely associated with competing schools in economics. We may classify them into three groups: the equilibrium school led by neo-classical economics, the disequilibrium school led by Keynesian, and the complexity school led by evolutionary economics. From mathematical perspective, both equilibrium and disequilibrium school are using linear models and static statistics, while complexity school is developing nonlinear and non-equilibrium approach. We will demonstrate how economic math play a key role in economic thinking and policy debate.

2.1 Single and Multiple Equilibriums in Linear and Nonlinear Demand Supply Curve

Neoclassical economics characterizes a self-stabilizing market by a single equilibrium state that is the cross point of the linear demand and supply curves (see Fig. 1a). In contrast, disequilibrium school describes market instability by multiple equilibriums, which are cross points of nonlinear demand and supply curves [7]. Their policy implications are quite explicit. Government interference is no need for single stable equilibrium, but crucial for avoiding bad equilibrium under multiple equilibrium situation.

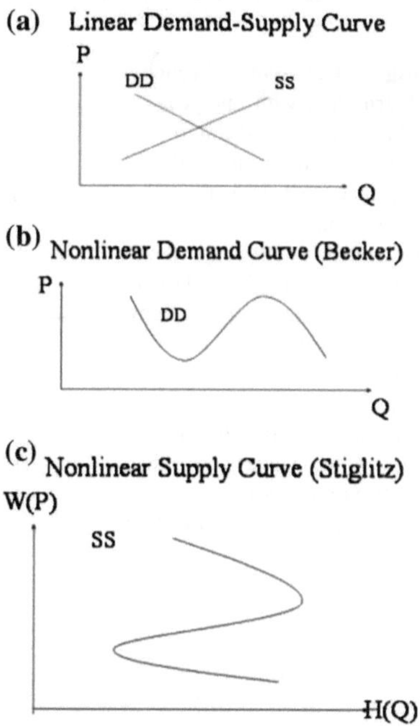

Fig. 1 Linear and nonlinear demand supply curves in micro. **a** Linear demand and supply with single equilibrium. **b** Nonlinear S-shaped demand with social interaction. **c** Nonlinear Z-shaped labor supply with subsistence and pleasure needs

Becker realized the possibility of S-shaped demand curve when social interaction plays an important role [8]. The herd behavior is observed from restaurant selection and market fads. Z-shaped labor supply curve implies increasing labor supply at low wage for survival needs, but reducing work at high wage for pleasure [9].

2.2 Large Deviation and Unimodular Distribution

How to understand the degree of market fluctuation? The probability distribution provides a useful tool in studying stochastic mechanism. The simplest distribution is the unimodular distribution with a single peak. The degree of market fluctuation is measured against the Gaussian distribution with finite mean and variance that is widely used in econometrics [10] and capital asset pricing model in finance theory [11]. If a random variable follows a Gaussian distribution, and its standard deviation is σ. The probability of larger than 3σ deviation from the mean is 0.3 %, and 5σ deviation of only 0.00006 %.

However, we often observe large market fluctuations or even crisis. Historical examples include the Black Monday in Oct. 19, 1987, Dot-com bubble in 1997–2000, and 2008 financial crisis. One possible explanation is a fat-tail distribution with single peak [12], such as the Cauchy distribution with infinite variance in probability theory. A special case is the power law studied in econophysics [13].

Both Gaussian and fat-tail belongs to the unimodular distribution with single peak. Their distributions are shown in Fig. 2.

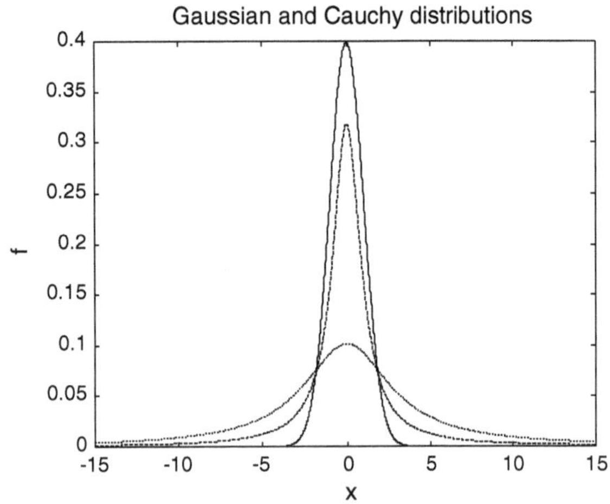

Fig. 2 Gaussian and cauchy distribution with one peak. The Gaussian distribution with zero mean and unit variance, N(0, 1), is the tallest in solid line. The Cauchy distribution with single peak but varying height is shown here for comparison. Cauchy (0, 1) distribution is in the middle in dashed line, and Cauchy (0, π) is the lowest and fattest distribution in dotted line. Clearly, fat-tail distribution has larger probability of large deviation

2.3 Bi-modular Distribution and Regime Switch in Social Psychology

What is the cause of large market fluctuation? Some economists blame irrationality behind the fat-tail distribution. Some economists observed that social psychology might create market fad and panic, which can be modeled by collective behavior in statistical mechanics. For example, the bi-modular distribution was discovered from empirical data in option prices [14]. One possible mechanism of polarized behavior is collective action studied in physics and social psychology [15]. Sudden regime switch or phase transition may occur between uni-modular and bi-modular distribution when field parameter changes across some threshold.

Here, we discuss two possible models in statistical mechanics. One is the Ising model of ferromagnetism; and another is population model of social interaction.

2.3.1 Ising Model of Social Behavior in Equilibrium Statistical Mechanics

The Ising model in equilibrium statistical mechanics was borrowed to study social psychology. Its phase transition from uni-modular to bi-modular distribution describes statistical features when a stable society turns into a divided society, like recent events in Ukraine and the Middle East [16]. Fig. 3 shows the three regimes in Ising model of public opinion.

2.3.2 Population Model with Social Interaction in Non-Equilibrium Statistical Mechanics

The problem of the Ising model is that its key parameter, the social temperature, has no operational definition in social system. A better alternative parameter is the intensity of social interaction in collective action. The three regimes of the probability distribution can be obtained by the master equation [17]. The U-shaped distribution in Fig. 4c is similar to the bi-modular distribution in Fig. 3c.

2.4 Business Cycles: White Noise, Persistent Cycles and Color Chaos

A more difficult issue in business cycle theory is how to explain the recurrent feature of business cycles that is widely observed from macro and financial indexes. The problem is: business cycles are not strictly periodic and not truly random. Their correlations are not short like random walk and have multiple frequencies that changing over time. Therefore, all kinds of math models are tried in business cycle

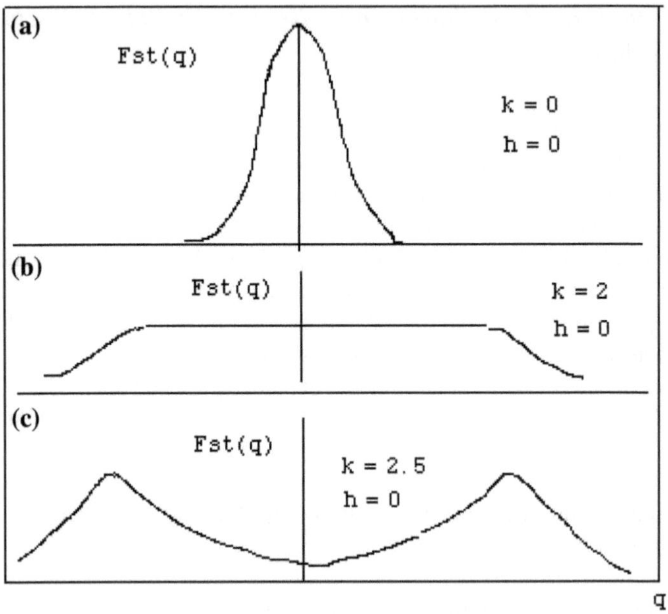

Fig. 3 The steady state of probability distribution function in the Ising Model of Collective Behavior with h = 0 (without central propaganda field). **a** Uni-modular distribution with low social stress (k = 0). Moderate stable behavior with weak interaction and high social temperature. **b** Marginal distribution at the phase transition with medium social stress (k = 2). Behavioral phase transition occurs between stable and unstable society induced by collective behavior. **c** Bi-modular distribution with high social stress (k = 2.5). The society splits into two opposing groups under low social temperature and strong social interactions in unstable society

theory, including deterministic, stochastic, linear and nonlinear models. We mainly discuss three types of economic models in terms of their base function, including white noise with short correlations, persistent cycles with long correlations, and color chaos model with erratic amplitude and narrow frequency band like biological clock.

Deterministic models are used by Keynesian economists for endogenous mechanism of business cycles, such as the case of the accelerator-multiplier model [18].

The stochastic models are used by the Frisch model of noise-driven cycles that attributes external shocks as the driving force of business fluctuations [19]. Since 1980s, the discovery of economic chaos [20] and the application of statistical mechanics [21] provide more advanced models for describing business cycles. We will show their main features in mathematical representation.

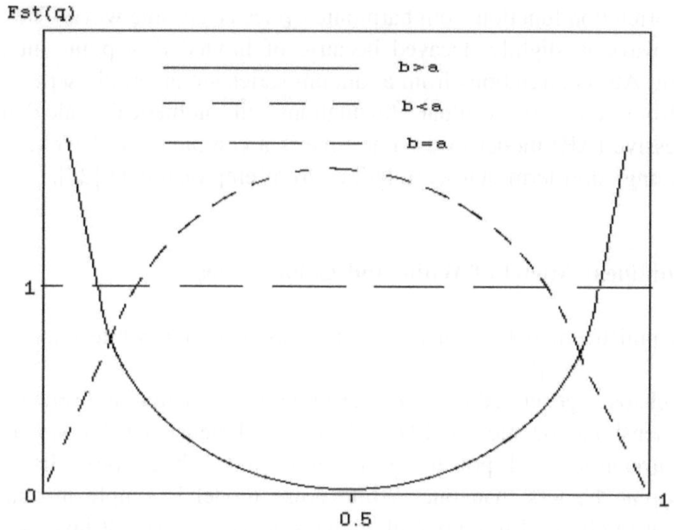

Fig. 4 The steady state of probability distribution function in socio-psychological model of collective choice. Here, "a" is the independent parameter; "b" is the interaction parameter. **a** Centered distribution with b < a (denoted by short *dashed curve*). It happens when independent decision rooted in individualistic orientation overcomes social pressure through mutual communication. **b** Horizontal flat distribution with b = a (denoted by *long dashed line*). Marginal case when individualistic orientation balances the social pressure. **c** Polarized distribution with b > a (denoted by *solid line*). It occurs when social pressure through mutual communication is stronger than independent judgment

Fig. 5 Numerical autocorrelations from time series generated by random noise and harmonic wave. The *solid line* is white noise. The *broken line* is a sine wave with period P = 1

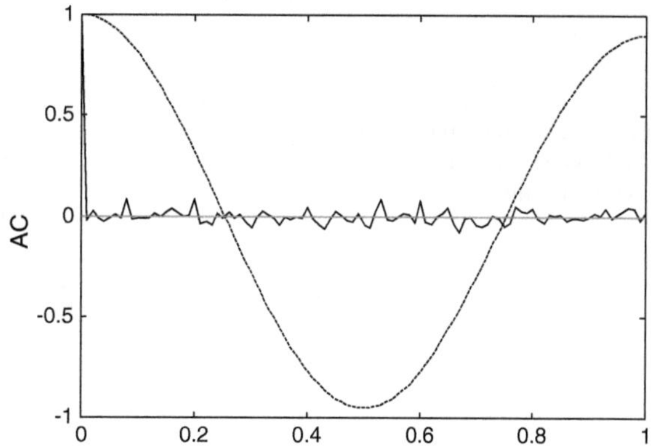

2.4.1 Linear Model of Harmonic Cycle and White Noise

Linear harmonic cycles with unique frequency are introduced in business cycle theory [22]. The auto-correlations from harmonic cycle and white noise are shown in Fig. 5.

Auto-correlation function from harmonic cycles is a cosine wave. The amplitude of cosine wave is slightly decayed because of limited data points in numerical experiment. Auto-correlations from a random series are an erratic series with rapid decade from one to residual fluctuations in numerical calculation. The auto-regressive (AR) model in discrete time is a combination of white noise term for simulating short-term auto-correlations from empirical data [23].

2.4.2 Nonlinear Model of White and Color Chaos

The deterministic model of chaos can be classified into white chaos and color chaos.

White chaos is generated by nonlinear difference equation in discrete-time, such as one-dimensional logistic map [24, 25] and two-dimensional Henon map [26, 27]. Its autocorrelations and power spectra look like white noise. Its correlation dimension can be less than one. White noise model is simple in mathematical analysis but rarely used in empirical analysis, since it needs intrinsic time unit.

Color chaos is generated by nonlinear differential equations in continuous-time, such as three-dimensional Lorenz model [28] and one-dimensional model with delay-differential model in biology [29] and economics [30]. Its autocorrelations looks like a decayed cosine wave, and its power spectra seem a combination of harmonic cycles and white noise. The correlation dimension is between one and two for 3D differential equations, and varying for delay-differential equation. We will show later that only color chaos is observed from empirical economic indexes.

The most visible feature of deterministic chaos is their phase portrait. The typical feature of color chaos is its spiral pattern in its phase portrait in Fig. 6.

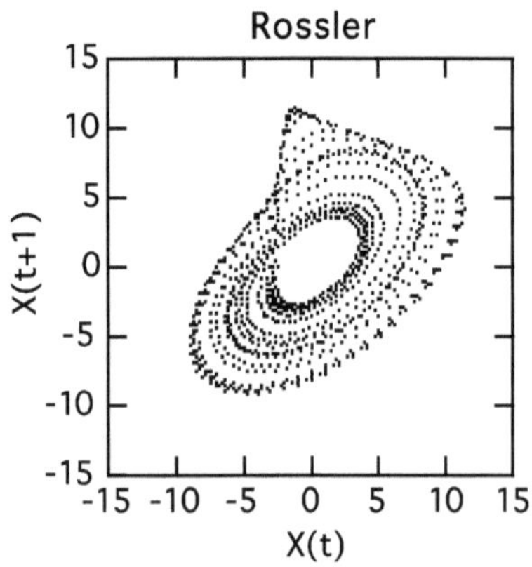

Fig. 6 The phase portrait of the Rössler color chaos in continuous time [31]

2.4.3 Fragile and Resilient Cycles in Linear and Nonlinear Oscillator

History shows the remarkable resilience of a market that experienced a series of wars and crises. The related issue is why the economy can recover from severe damage and out of equilibrium? Mathematically speaking, we may exam the regime stability under parameter change.

One major weakness of the linear oscillator model is that the regime of periodic cycle is fragile or marginally stable under changing parameter. Only nonlinear oscillator model is capable of generating resilient cycles within a finite area under changing parameters. The typical example of linear models is the Samuelson model of multiplier-accelerator [32]. In Fig. 14a, periodic solution PO only exists along the borderline between damped oscillation (DO) and explosive oscillation (EO).

Figure 7b, c is the parameter space of the nonlinear dynamic model of soft-bouncing oscillator [33]. Its nonlinear periodic solution P1 (period one, the limit cycle), P2 (period 2), P3 (period 3), and CH (chaotic solution) are structurally stable when parameter changes are within the CP area. The phase transition or regime switch occurs when parameter change is cross the boundary of CH or CP.

Linear stochastic models have similar problem like linear deterministic models. For example, the so-called unit root solution occurs only at the borderline of the unit root [34]. If a small parameter change leads to cross the unit circle, the stochastic solution will fall into damped (inside the unit circle) or explosive (outside the unit circle) solution.

2.5 Logistic Wavelet, Metabolic Growth and Schumpeter's Creative Destruction

Any living system has a life cycle that can be described by a wavelet with finite life, which is a better math representation than harmonic wave with infinite life or white noise with zero life.

The origin of wavelet representation in economics can be traced to resource limit in market-share competition. The ecological constraint is the major source of economic nonlinearity [35]. The simplest model of resource limit in theoretical ecology is the logistic growth. Species competition in ecology can be used as technology competition model. A well-known example is the Lotka-Volterra model with two species. Figure 8 shows the outcome of two competing species with different carrying capacity.

The logistic wavelet with a finite life is a simple nonlinear representation for technology life cycles. Schumpeter's long waves and creative destruction can be described by a sequence of logistic wavelets in a technology competition model [36].

A numerical solution of competition equation is shown in Fig. 8. Without technology 2's competition, the growth path of technology (species) 1 (on the left)

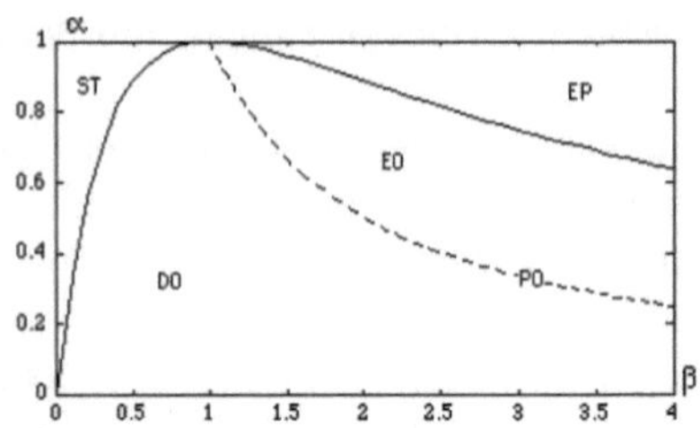

(a) Stability pattern of Samuelson model in parameter space (1939). Here, ST denotes the steady state; DO, damped oscillation; EO, explosive oscillation; EP, explosive solution; PO, linear periodic oscillation.

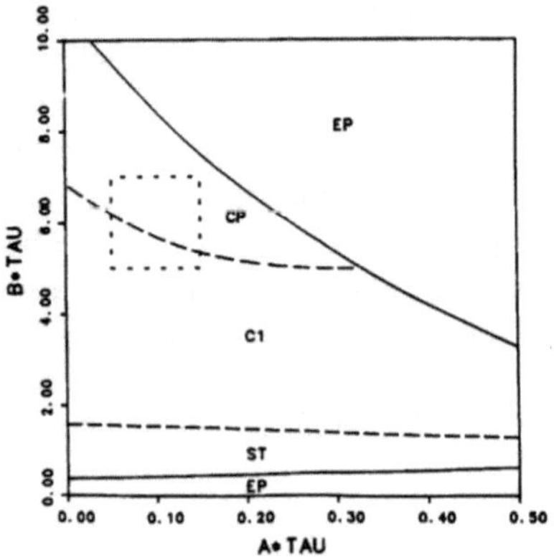

(b) Parameter space for soft-bouncing oscillator. ST denotes the steady state. CP is the complex regime including multi-periodic states C1, C2, C3, etc.

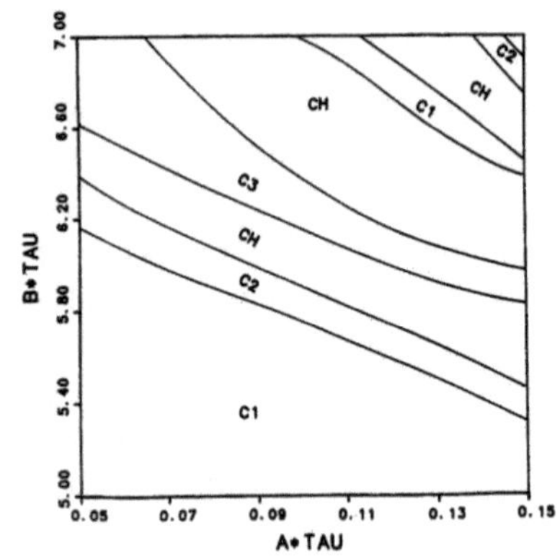

(c) The expanded regime in (11b). C1, C2, C3 are limit cycles of period one, period two, and period three respectively; CH, the chaos mode in continuous time.

Fig. 7 Structural stability in parameter space. **a** Periodic solution PO is only marginally stable at the borderline. **b** Complex and chaotic regime is structurally stable within the area of CP. The complex regime CP in **b** is enlarged in CH in **c** that consists of alternative zones of limit cycles and chaos

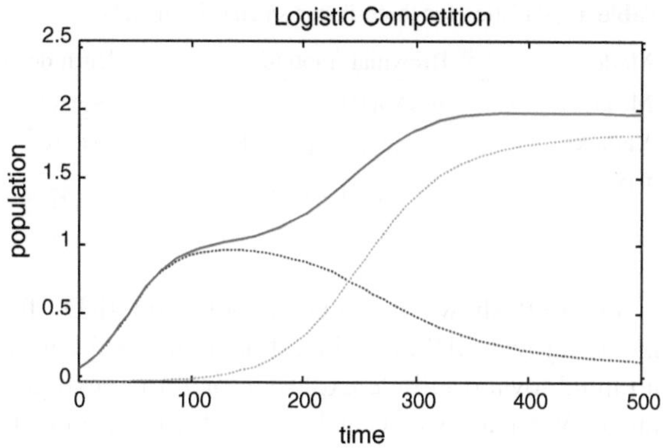

Fig. 8 Staged economic growth is characterized by the envelop of the sum of competing logistic wavelets that mimics the observed pattern of a time series from macroeconomic indexes

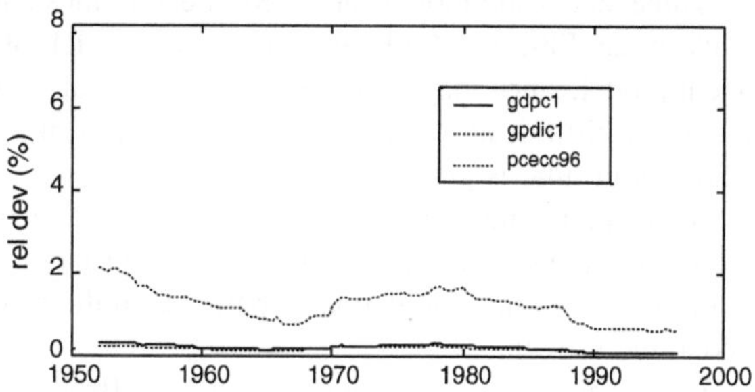

Fig. 9 The RDs of GDPC1 (US real GDP), GPDIC1 (real investment), and PCECC96 (real consumption) for the US quarterly series (1947–2001). N = 220. Moving time window is 10 years. Displayed patterns were observed through the HP filter

would be a S-shaped logistic curve (on the right). However, the realized output of technology 1 resulting from competition with technology (species) 2 looks like an asymmetric bell curve. We call it the logistic wavelet, which is a result from the competition of new technology. Clearly, logistic wavelet is a better representation for technology progress than random shocks in RBC (real business cycle) literature, since the technology competition model described Adam Smith's idea of market-share competition [37]. Schumpeter's creative destruction can be explained by over-capacity under technology competition, since old and new technology market-share are below their full capacity. The so-called "insufficient aggregate demand" in Keynesian economics could be resulted from "excess capacity" caused by technology competition for market share.

2.6 Persistent Fluctuation and Population Dynamics

A new measurement of relative deviation (RD) reveals the relation between persistent fluctuations and population dynamics. Economic statistics often measure mean and variance separately. For positive variables, such as price and population, RD is the ratio of standard deviation to its mean [38]. RD is an important indicator for persistent fluctuations (see Fig. 9).

Table 1 Statistics for three linear stochastic models

Model	Brownian motion	Birth-death	Random-walk
Mean	$\sim \exp(rt)$	$\sim \exp(rt)$	~t
Variance	$\sim \exp(2rt)\{e^{\sigma^2 t} - 1\}$	$\sim e^{rt}(e^{rt} - 1)$	~t
RD	$\sim e^{\frac{\sigma^2}{2}t}\sqrt{(1 - e^{-t\sigma^2})}$	$\sim \frac{1}{\sqrt{N_0}}$	$\sim \frac{1}{\sqrt{t}}$

Figure 9 shows the relative deviation (RD) from three US macro indexes, including real GDP, real investment, and real consumption. All three curves have no trend toward infinite explosion or zero convergence, only oscillate within finite range. We may ask what kind of stochastic models has the feature of persistent fluctuations.

Three stochastic models are used in economics and finance theory. The random walk model [39] and the Brownian motion model [40] belong to the representative agent model, since they describe the trajectory of only one particle. The population model of birth-death process is also introduced in physics and finance [41]. Their statistics in time is given in Table 1.

In this table, the first moment is the mean, the second moment is variance, RD is the ratio of the standard deviation (the square root of variance) to its mean. Here, N_0 is the size of initial population of particles in the birth-death process and r > 0 for economic growth.

From Table 1, we can see that the RD of Brownian motion is explosive in time, RD of random walk is damping. Only the RD of the birth-death process tends to a constant. This is a critical criterion in choosing proper math model in finance theory [42].

Clearly, the Brownian motion and random walk model are not proper for modeling macro dynamics in the long-term. Only the birth-death process is qualified for both macro and financial dynamics with persistent RD. We found that Brownian motion model is explosive in 2005. We speculated that the collapse of the derivative market in 2008 financial crisis might be caused by the theoretical flaw in option pricing model, since interest swap pricing model is also based on the Brownian motion. We suggest that option pricing should be based on the birth-death process [43, 44].

2.7 *The Frisch Model of Noise-Driven Cycles: A Math Delusion for Self-stabilizing Market or an Economic Fallacy as a Perpetual Motion Machine?*

The Frisch model of noise-driven cycles plays a central role in equilibrium theory of business cycle and econometrics. In mathematical construction, it is a mixed model of damped harmonic cycles and persistent white noise [45]. A harmonic oscillator with friction would automatically tend to stop. This dynamic feature is used for characterizing a self-stabilizing market. The problem is how to explain persistence of business cycles in history. Frisch suggested that harmonic cycles could be maintained by persistent shocks. Frisch made the claim during the Great Depression in 1933. If the Frisch model is true, the Frisch model will save the liberal belief in a self-stabilizing market, since persistent business cycles are not generated by internal market stability but external random shocks. Frisch shared the first Nobel Prize in economics for this model in 1969. The noise-driven model is also behind the works by Lucas and RBC school.

However, physicists already knew before Frisch that the demped harmonic cycles could not be kept alive by random shocks [46], since its amplitude would decay exponentially [47]. We calibrate the Frisch model by the US data. We found that American business cycles would only last about 4–10 years [48]. In fact, the NBER recorded US business cycles since 1854, more than 160 years ago.

The Frisch conjecture implies a perpetual motion machine of the second kind. It is a working machine powered by random thermal fluctuations, which is a heat engine with single temperature source. This engine could not do any work, since any heat engine must burn fuel at high temperature and release waste heat at the low temperature according to the second law of thermodynamics.

Frisch claimed that he had already solved the analytical problem and that this would soon be published. His promised paper was advertised three times under the category "papers to appear in early issues" in 1933, but it never appeared in *Econometrica*, where Frisch served as the editor. Frisch did not mention a word about his prize-winning model in his Nobel speech in 1969 (Frisch 1981). Obviously, Frisch quietly abandoned his model since 1934, but never admitted his mistake in public. This story should be a wake-up call to equilibrium economics: there is only one step between economic belief and mathematical delusion.

3 Empirical Tests of Competing Economic Math

Empirical tests of competing math models in economics is more difficult than that in physics. There are two fundamental problems to be solved in conducting numerical tests of competing economic models.

First, many economic time series have a growing trend, since economy is an open system with increasing energy consumption. However, existing models of business cycles are stationary models. We need a proper mapping to transform non-stationary time series into stationary series, which is a similar issue in finding a preferred observation reference in planet motion. This is the Copernicus problem in economics.

Second, economic data is raw data that contain strong noise, which is more difficult for signal processing comparing to low noise data from controlled experiments in natural science. Conventional band-pass filter is not sufficient to separate signal from noise. We need more advanced tools in analyzing non-stationary time series.

In this section, we will introduce two math tools in solving these two problems: the HP filter for defining smooth trend series and the time-frequency analysis in time-frequency two-dimensional space.

3.1 Observation Reference and the Copernicus Problem in Economics

Physics revolution started with the Copernicus, who changed the geocentric system by Ptolemy to heliocentric system for studies of planet motion. Economic analysis faces a similar choice of preferred observation reference.

Competing economic schools differ in their scope of time windows in analyzing economic time series.

Equilibrium school and econometrics believe that market is capable in self-adjustment towards equilibrium, so that they choose a short-term time window to obtain a random image of market movements. Their tool is calculating the rate of changes within a time unit from a growing time series, such as the daily or yearly returns of stock prices or GDP series. Mathematically speaking, it is equivalent to the first difference of a logarithmic time series, which is called the FD (first differencing) filter in econometrics. The FD de-trended time series look random. This is the empirical foundation of the so-called efficient market. An alternative mapping is taking a log-linear trend, which depends on the end points or the length of time series. The HP (Hodrick-Prescott) filter is a compromise between the FD and the log-linear detrending (LLD) filter. The HP filter defines a nonlinear smooth trend, which corresponds to a medium time window in the typical length of NBER business cycles about 2 to 10 years [49]. The HP filter was discussed by Von Neumann in analyzing time series with changing mean [50]. We found out that the FD series look random but the HP cycles reveal nonlinear patterns and varying frequency. We may compare their results in Fig. 13.

From the length of the first zero autocorrelation in Fig. 10c, we may estimate the length of the period from detrended series. FD cycle is 0.7 year, HPc 3 years, LLD 29 years. Clearly, HP cycles are the best indicator to characterize the US business cycles.

3.2 Filter Design in Signal Processing

Signal processing is a technology field, which rapidly developed since 1950s. Scientists make a great effort to reduce noise level and enhance signal for getting better information. Band-pass filter is widely used in empirical analysis. Only one

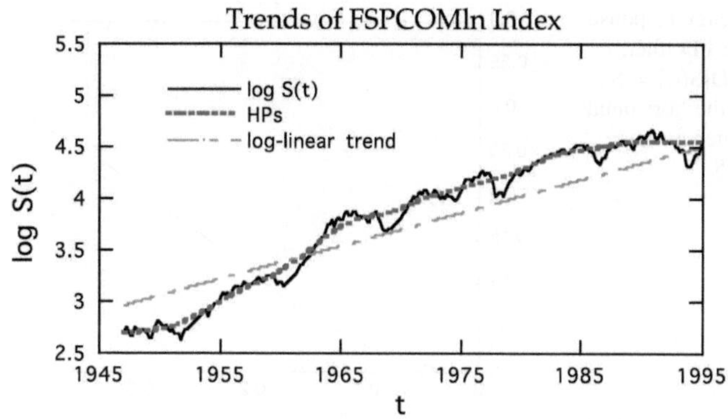

(a). HP trend and LLD (log-linear) trend for X(t) {=log S(t) }. LLDc cycles are residuals from log-linear trend.

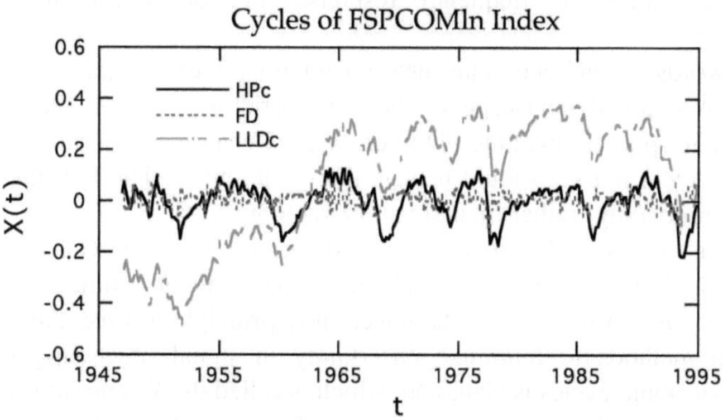

(b). Cycles from competing detrending filters.

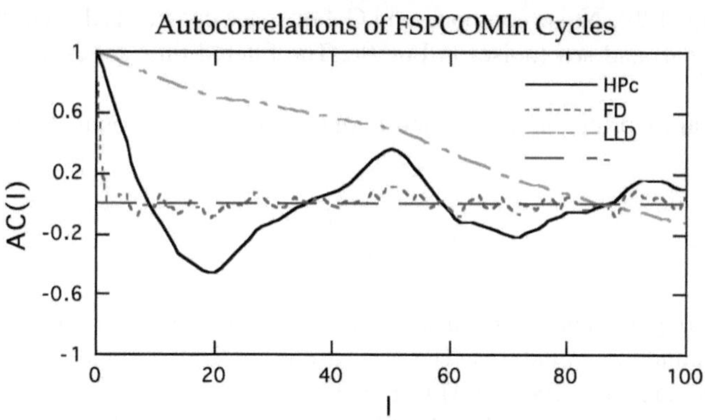

(c). Autocorrelations of detrended series. The length of correlations is the longest from LLD, medium from HP, and shortest from FD.

Fig. 10 Varying correlated length from FD, HP, and LLD detrending filtered from the logarithmic FSPCOM (S&P 500) monthly series (1947–92). N = 552

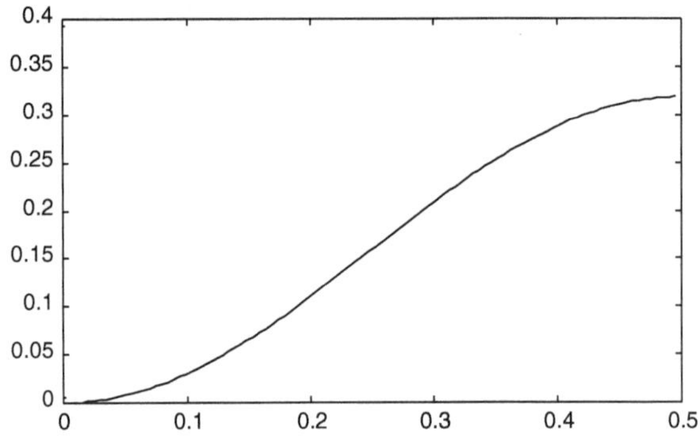

Fig. 11 Frequency response function for the FD filter. Here, X(t) = FD[S(t)] = S (t + 1) − S(t), the horizontal axis is the frequency range from zero to 0.5

discipline made an odd choice: The FD filter in econometrics is used for amplifying high frequency noise. Its frequency response function reveals its secret in Fig. 11, [51].

In other words, econometricians use a whitening looking glass in empirical observation. A colorful landscape in the real world would look like randomly distributed white spots through the FD filter. Why? Because equilibrium economics asserts that efficient market should behave like white noise. The FD filter plays a role of illusion creating, rather than signal extracting.

In contrast, we developed a more advanced filter in two-dimensional time-frequency Gabor space for analyzing non-stationary time series [52]. According to quantum mechanics, the uncertainty principle in time and frequency indicates two methods to minimize uncertainty in signal processing when the envelope of harmonic cycles is Gaussian, which is called the Wigner transform. We can project the original time series onto a two-dimensional time-frequency Gabor discrete lattice space (see Fig. 12a), and using Wigner function as base function in every grid (Fig. 12b). Noise component in Gabor space can be easily separated like mountain (signal) and sea (noise) in Fig. 9c. The filtered time series are shown in Fig. 13.

The uncertainty principle in time and frequency is the very foundation of signal processing:

$$\Delta f\, \Delta t \geq \frac{1}{4\pi} \tag{1}$$

We find strong evidence of persistent cycles and color chaos through HP cycles from financial and macro indexes. The best case is the S&P 500 index (its monthly data series are named as FSPCOM) in Fig. 13.

From Fig. 13, we can see that the filtered $S_g(t)$ series explain 70 % of variance from the original $S_o(t)$ series, which is the cyclic component through the HP filter. Their cross-correlation is 0.847, which is very high in computational experiment. The correlation dimension is the numerical measurement of fractal

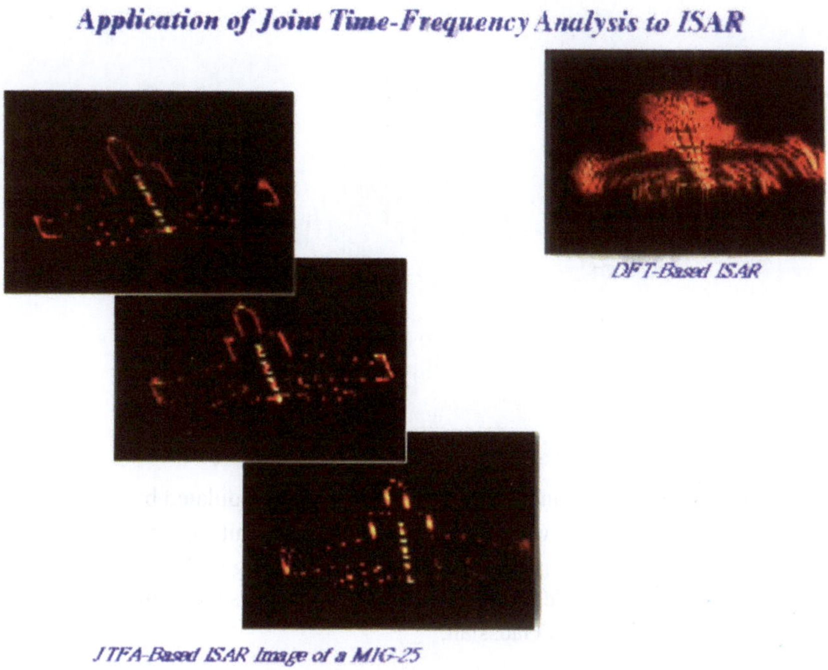

(a) Non -stationary (left picture) vs. stationary (right picture) time series analysis of a moving object (MIG-25 fighter plane).

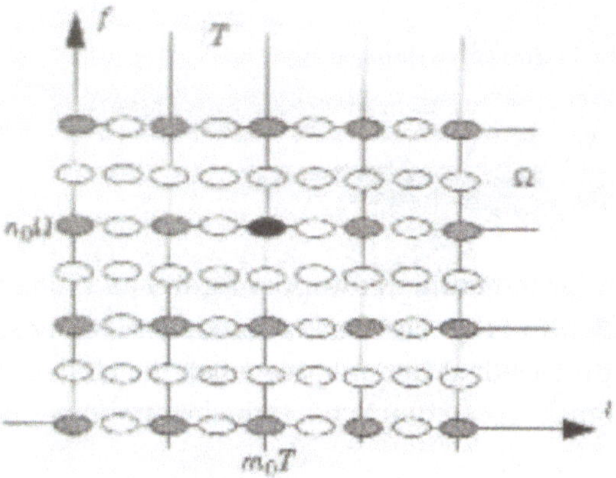

(b) Gabor Lattice Space in Time and Frequency

Fig. 12 Construction and application of the time-variant filter in Gabor space

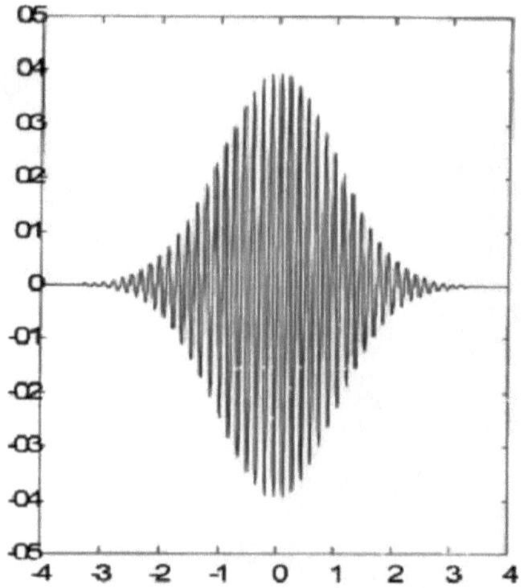

(c) Wigner base function is a harmonic wave modulated by a Gaussian envelope, which has infinite span but finite standard deviation for analyzing local dynamics. According to the uncertainty principle, the wave uncertainty reaches the minimum when its envelope is Gaussian.

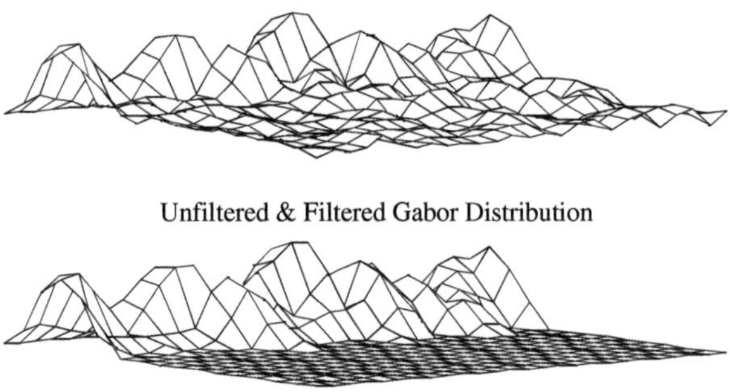

(d). The Gabor distribution for the unfiltered (upper figure) and filtered (lower figure) data of FSPCOMln HPc. The noise component (below sea level) was removed in two-dimensional Gabor space when the HPc series were decomposed by Wigner base functions, which minimize the uncertainty.

Fig. 12 (continued)

dimension, which are 2.5 for S&P 500 monthly index. This is solid evidence that stock price movements can be characterized by nonlinear color chaos. Its average period is 3 years. Therefore, our analysis supports Schumpeter theory of business

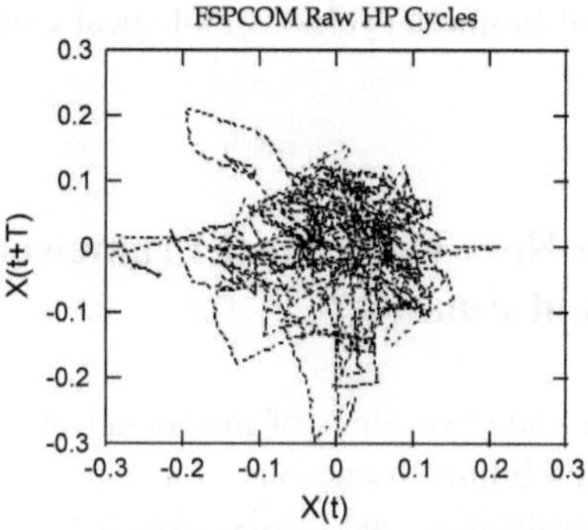

(a) Noisy image of unfiltered data

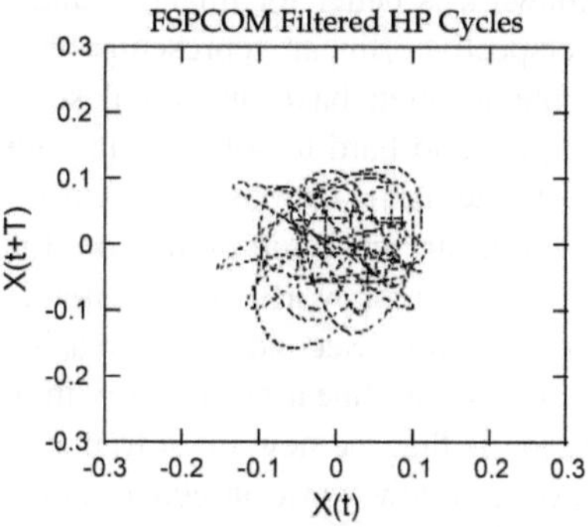

(b) Complex spiral of filtered data

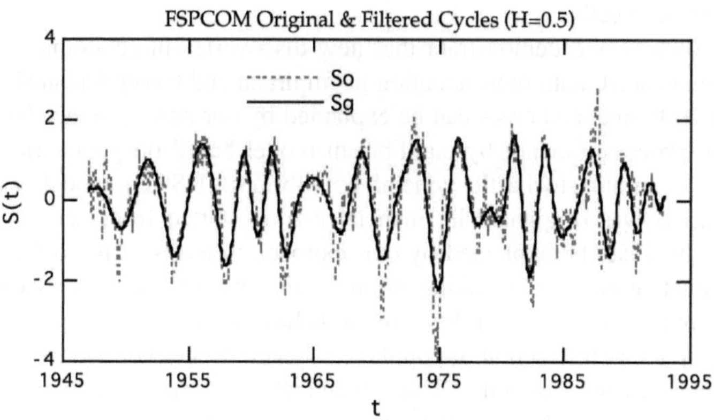

(c). The original $S_O(t)$ and filtered $S_g(t)$ from FSPCOM HPc.

Fig. 13 Comparison of the original and filtered time series of FSPCOMln HP Cycles

cycles [53]. He considered business cycles as biological clock, a living feature of economic organism.

4 Why We Need a New Theoretical Framework for Economics and Finance?

The above discussions provide two kinds of mathematical representations. One is based on the linear-equilibrium framework for an efficient market that is self-stabilizing without instability and crisis. Another is based on nonlinear non-equilibrium framework for economic complexity with cycles and chaos. Now the question is which framework is better for financial and economic research?

From mathematical perspective, linear representation is simpler and more beautiful, since linear problems often have analytical solutions, while nonlinear representation is more complex and hard to solve. Many nonlinear problems have no analytical solution and numerical solutions are often controversial. However, physicists prefer to use nonlinear math representation for three reasons: First, physicists have to deal with the real world that is nonlinear, non-equilibrium, non-stationary, and complex in nature. Second, the increasing computational power of computers is capable of solving nonlinear problems with desired precision. Third and the most important reason is that the new math framework based on nonlinear non-equilibrium approach open a new world in economics and finance. Similar to non-Euclidean geometry in relativity theory, complexity science reveals new evidence of economic structure and historical changes, which is not known in linear equilibrium perspective.

In this section, we demonstrate that new discoveries in economic observation need new advanced math representation in empirical and theoretical analysis. Three lessons in 2008 financial crisis can be explained by our new approach based on the birth-death process, but not by equilibrium model based on geometric Brownian motion or dynamic stochastic general equilibrium (DSGE) model. First, 2008 financial crisis was originated in financial market started from derivative market collapse, which can be explained by our approach in terms of meso foundation of business fluctuations and breaking point in the master equation. Second, stock market bubbles often caused by animal behavior or herd action that can be described by social interaction in population dynamics, but not representative agent model in mainstream economics. Third, financial crisis implies sudden changes and regime switch that is beyond the scope of static distribution. In contrast, the birth-death process is a continuous-time model that can be solved by the master differential equation. Its probability distribution is varying over time that is compatible with historical observation. Its nonlinear solution is capable of characterizing phase transition and regime switch.

Let us see how advanced math representation opens new windows in economic analysis.

4.1 *The Principle of Large Numbers and the Discovery of Meso Foundation*

Structure analysis plays an important role in physics and biology. However, structure is missing in macroeconomics. The so-called microfoundations theory simply asserts that macro dynamics should follow the same formulation in microeconomics. We will study the micro-macro relation in business cycle theory.

One fundamental issue in macro and finance theory is the origin of business cycles and the cause of the Great Depression. Lucas claimed that business cycles or even the Great Depression could be explained by workers' choices between work and leisure, which is called the micro-foundations theory of (macro) business cycles. How can we discover the micro-macro relation in stochastic mechanism? Schrödinger proposed a simple math that reveals the relation between the number of micro elements and the degree of aggregate fluctuations [54]. We define the relative deviation (RD) as the ratio of the standard deviation to its mean when the underlying variable has only positive value, such as price and volume.

$$RD = \frac{STD(S_N)}{\sqrt{N}} \tag{2}$$

Here, RD stands for relative deviation for positive variable, STD is standard deviation, which is the square root of the variance of a variable S with N elements: $S_N = X_1 + X_2 + \cdots + X_N$.

The idea is quite simple. The more element number N at the micro level, the less will be the aggregate fluctuation at the macro level, since independent fluctuations at the micro level would largely cancel out each other. We call this relation as the principle of large numbers. We extend this relation from static system to the population dynamics of the birth–death process [55]. We first calculate RD from an economic index through the HP filter. Then, we estimate the effective micro number N. The result is given in Table 2, which can be used for diagnosing financial crisis [56].

In comparison, the number of households, corporations and public companies and the potential RD generated by them are given in Table 3.

From Tables 2 and 3, household fluctuations may contribute only about 5 % of fluctuations in real gross domestic product (GDP) and less than 1 % in real investment; and small firms can contribute 50 % of fluctuations in real GDP or 8 % in real investment. In contrast, public companies can generate about 60 % of

Table 2 Relative Deviation (RD) and Effective Number (N) for macro and finance indexes

Item	RD (%)	N
Real personal consumption	0.15	800,000
Real GDP	0.2	500,000
Real private investment	1.2	10, 000
Dow Jones Industrial (1928–2009)	1.4	9,000
S&P 500 index (1947–2009)	1.6	5,000
NASDAQ (1971–2009)	2.0	3,000
Japan–US exchange rate (1971–2009)	6.1	300
US–Euro exchange rate (1999–2009)	4.9	400
Texas crude oil price (1978–2008)	5.3	400

Table 3 Numbers of households and firms in US (1980)

Micro-agents	Households	Corporations[a]	Public companies
N	80 700 000	2 900 000	20 000
RD (%)	0.01	0.1	0.7

[a]Here, we count only those corporations with more than $100 000 in assets

aggregate fluctuations in real investment. Clearly, there are very weak 'micro-foundations' but strong evidence of a 'meso-foundation' in macroeconomic fluctuations.

In another words, large macro fluctuations in macro and finance can only be generated by fluctuations at the meso (finance) level, not the micro level from households or small firms. Extremely large fluctuations in commodity and currency market can only be caused by financial oligarchs. This is the root of 2008 financial crisis [57]. Therefore, competition policy is more effective than monetary and fiscal policy in stabilizing macro economy and financial market. Therefore, we strongly recommend that breaking-up financial oligarchs is the key to prevent the next financial crisis. This is the most important lesson that was learned from our new theoretical framework of the birth-death process.

Our approach finds strong evidence of meso (finance and industrial organization) structure from macro and finance indexes. Our three-level system of micro-meso-macro is better than the two-level system of micro and macro in Keynesian economics in studies of structural foundation of business cycles and crisis.

4.2 *Transition Probability in Calm and Turbulent Market: A Diagnosis of 2008 Financial Crisis*

Efficient market theory in finance simply rules out the possibility of financial crisis. Static model of market instability is not capable in quantitative analysis of financial

Fig. 14 Transition Probability for Calm (1950–1980) and Turbulent (1980–2010) Market Regimes. The horizontal axis is the price level of the S&P 500 daily index. The vertical axis is the transition probability at varying price level. Data source is S&P 500 daily close prices

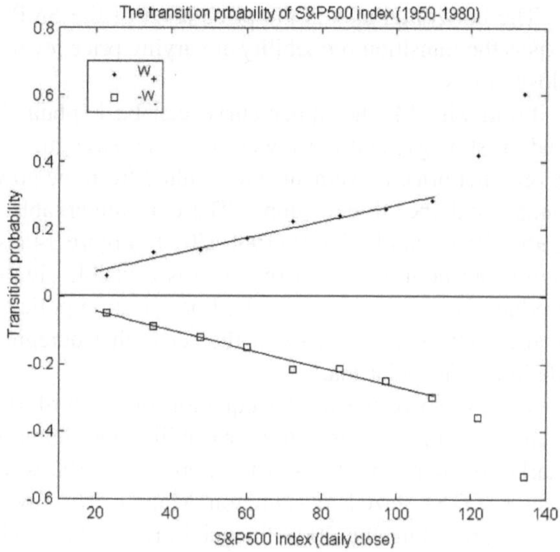

(a) Transition probability in (1950-1980).

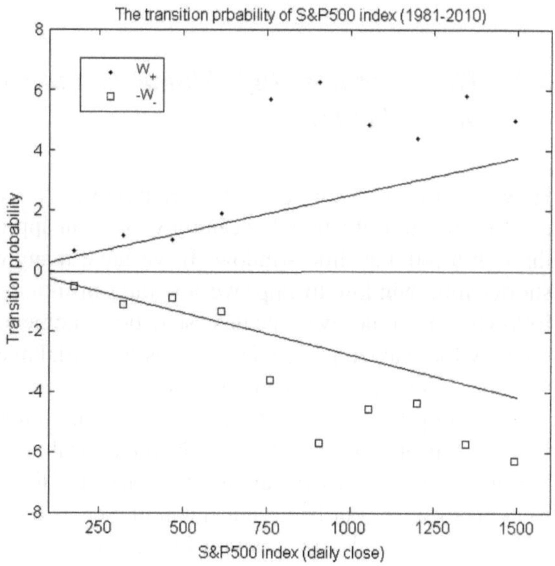

(b) Transition Probability in (1980-2010).

indexes. We apply non-equilibrium statistical mechanics to the birth-death process in analyzing the S&P 500 daily indexes from 1950 to 2010 [58]. The state-dependent transition probability is shown in Fig. 14.

The horizontal axis is the price level of the S&P 500 daily index. The vertical axis is the transition probability at varying price level. Data source is S&P 500 daily close prices.

From Fig. 14, the upper curve can be explained by the "strength" with buy trading strategy, and the lower curve the strength with sell trading strategy. Intuitively, net price movements are resulted from the power balance between the "Bull camp" and the "Bear camp". There is remarkable difference between Period I (1950–1980) and Period II (1980–2010). Figure 14a is smoother than Fig. 14b. The significant nonlinearity in Fig. 14b is a visible sign of turbulent market that may produce financial crisis. Clearly, liberalization policy in Period II is closely related to the 2008 financial crisis in the sense that deregulation stimulated excess speculation in financial market.

We can solve the master equation of the birth-death process and find out the break point of the distribution probability. Our numerical solution indicates that the market breakdown occurs at the Sept. 25, 2008, when the Office of Thrift Supervision (OTS) seized Washington Mutual. This event was the peak from chain events preceding the 2008 financial crisis. The stock market went to panic since 26-Sep-2008. Our result is well compatible with historical timeline.

4.3 Time-Varying High Moments and Crisis Warning in Market Dynamics

How to observe an evolving economy? If economy is time-varying in non-equilibrium situation, we can only make an approximation by local equilibrium through a moving time window. If we have many data points, we may choose a shorter time window to improve foresting in monitoring market instability. In the following example, we calculate statistical moments through a moving quarterly window for analyzing daily Dow Jones Industrial average (DJI) index that provides valuable information on coming crisis.

The sub-prime crisis in the U.S. reveals the limitation of diversification strategy based on mean-variance analysis. Neo-classical economics is mainly interested in the first two moments (mean and variance) for the whole series based on equilibrium perspective. For better understanding crisis dynamics, we exam high moments before, during, and after crisis in Fig. 15, [59].

A regime switch and a turning point can be observed using a high moment representation and time-dependent transition probability. Financial instability is visible by dramatically increasing 3rd–5th moments one-quarter before and during the crisis. The sudden rising high moments provide effective warning signals of a regime-switch or a coming crisis.

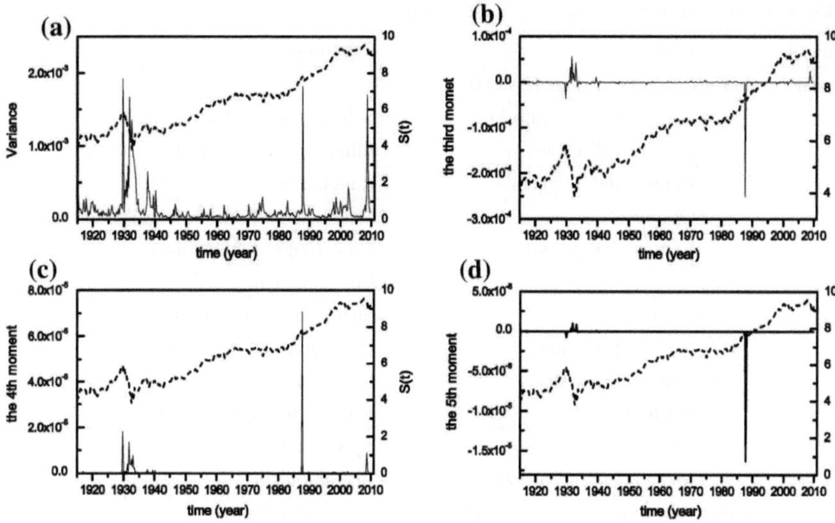

Fig. 15 The quarterly moments (*solid lines*) of the Dow-Jones Industrial Average (DJI) index. The original S(t) (*dashed lines*) is the natural logarithmic daily close price series. Each point in the *solid line* is calculated with a moving time window; its width is one quarter. Plots **a**, **b**, **c** and **d** correspond to 2nd, 3rd, 4th and 5th moment, respectively. The magnitudes of each moment representation are 10^{-5} for variance, 10^{-8} for 3rd moment, 10^{-9} for 4th moment, and 10^{-11} for 5th moment. The daily data were from 2-Jan-1900 to 1-Sep-2010 with 27724 data points. Here we choose $\sigma_0^2 \sim 10^{-5}$ as the normal level. We would consider high moments when they reach the level of $10^{-1}\sigma_0^2$ or higher

5 Math Selection and Philosophical Preference

Now, we have two systems of mathematical representations in economics and finance. Their main differences are given in Table 4.

Neoclassical economics constructed a mathematical utopian of the market. The market is self-stabilizing without internal instability, since it has unique stable equilibrium. There is no resource limit to utility function and exponential growth. Therefore, human nature is greedy with unlimited want. Rational economic man makes independent decision without social interaction. The representative agent extremely simplifies economic math to a single-body problem, such as random walk, Brownian motion, and stochastic dynamic general equilibrium (SDGE) model. Resource allocation problem can be solved by optimization with complete market and perfect information without open competition and technology change. Price is the single variable that is capable of determining output and distribution. Business cycles and market fluctuations are explained by external shocks, so that there is little need for market regulation and government intervention. Reductionism and methodological individualism is justified by the linear approach of economic math since the whole equals the sum of parts.

Table 4 Competing ideas in economics and math

Subject	Math	Economics
Linear-equilibrium paradigm (neoclassical)		
	Single stable equilibrium	Self-stabilizing, laissez-fair
	Gaussian distribution	Efficient market, small deviation
	Representative agent	Rational man
	Optimization	Complete market, closed sys.
	Exponential growth	No ecological constraints
		Consumerism, greedy
	White noise	Short correlation, no history
	Brownian motion	Homogeneous behavior
	Regression analysis	Integrable system
	FD filter	Short-term view, whitening signal
	Mean reversion	System convergence
	Mechanical philosophy	Reductionism
Nonlinear-nonequilibrium paradigm (complex-evolutionary economics)		
	Multiple equilibriums	Multiple regimes, instability, crisis
	Time-varying distribution	Calm & turbulent market
	Population dynamics	Social species within ecology
	Logistic wavelets	Tech. competition for resource
		Creative destruction
	Complex dynamics	Complex evolution in open system
	Color chaos & wavelets	Life cycle, biological clock,
		Medium-correlation
		History (path-dependency)
	Birth-death process	Herd behavior, social interaction
	Time-frequency analysis	Non-integrable, non-stationary sys.
	HP filter	Medium-term view, trend-cycle
	Bifurcation tree	System divergence
	Biological philosophy	Holistic + structure = complexity

The linear demand-supply curve is powerful in creating a market belief among public. That is why economic math is dominating in ivory tower economics. The only trouble for this linear-equilibrium world is a lacking understanding of recurrent crisis. One possible excuse is the existence of fat-tailed distribution. Even we have a big chance of large deviation, we still have no policy to deal with the bad luck, when we have a static unimodular probability distribution.

Complexity science and evolutionary economics discovered a complex world by nonlinear dynamics and Nonequilibrium physics. The real market has several regimes including calm and turbulent market, which can be described by multiple equilibrium and time-varying multi-modular distribution. Resource constraints shape nonlinear dynamics in economic decision and social interaction. Internal instability is driven by technology metabolism and social behavior. Human beings are social animals in nature. Macro and financial market must consider population dynamics and collective behavior. The representative agent model is misleading since collective behavior plays an important role in market fads and panic. Economic dynamics is complex since structure and history plays important role in economic movements. Life cycle in technology can be described by wavelets. Business cycles can be characterized by color chaos model of biological clock. Financial crisis can be better understood by time-frequency analysis and time-varying transition probability. Competition policy and leverage regulation provides new tools in market management in addition to monetary and fiscal policy.

Complexity science and evolutionary perspective is developing an alternative framework including micro, macro, finance, and institutional economics. Nonlinear math needs more understanding in history and social structure. The rapid development in big data and computer power paves the numerical foundation of the nonlinear-nonequilibrium economic math.

Now economists are facing a historical choice: how to select a proper math system for economics and finance.

There are three possible choices.

For strong believers in invisible hand and laissez-faire policy, existing neo-classical model is simple enough to portrait a self-stabilizing market, such as linear demand-supply curve, convex function in optimization, and noise-driven business cycle theory. The problem is that mainstream economics have to dodge difficult issues, such as excess volatility, herd behavior, and financial crisis.

For applied mathematicians in economic research, they often face a dilemma in choosing mathematical simplicity or computational complexity. Linear models may have analytical solution that is beautiful for academic publication, while nonlinear models may be more realistic for policy practice but complicated for academic research. Currently, mainstream economic journals are falling behind science journals in advancing new math.

For empirical scientists doing economic research, there is an increasing trend in applying complexity science and non-equilibrium physics to economics and finance. New fields, such as the self-organization science, complex systems, and econophysics, are rapidly developed since 1980s.

Some economists may have doubt about physics application in economics. Our answer is: the proof of the pudding is in the eating. We already witness the success of physics application in medicine, including X-ray, ultrasound, laser, cardiograph, and DNA analysis. Economics more closely resembles medicine than theology. Complexity science and nonequilibrium physics has many applications in studying economics and finance, since the economy is an open system that is nonlinear and non-equilibrium in nature [60].

In philosophical perspective, the shift in math representation may facilitate a paradigm change [61]. The dominance of linear—equilibrium paradigm in mainstream economics sounds like theoretical theology in economics when it fails to address contemporary challenges, such as economic crisis and global warming [62]. Unlimited greed in human nature is not compatible to conservation law in physics and ecological constraints in earth [63].

The prevalence of noise-driven model in business cycle theory and regression analysis in econometrics is rooted in positive economics [64] or pragmatism in econometrics [65]. Now we have better knowledge why positive economics as well as econometrics fails to make economic prediction. Economic time series are non-stationary driven by structural changes in economy. There is little chance that model parameters may remain constant in time. That is why regression analysis is not reliable in economic research.

Mathematics is part of science knowledge that is constantly expanding by human experiences. Theoretical framework is developed from special knowledge to general knowledge, and accordingly, from simple to complex. For example, math developed from integer to fraction, from rational to irrational, from real to complex number, and from Euclidean to non-Euclidean geometry. Economic concepts will face similar changes in physics: from close to open systems, from equilibrium to non-equilibrium, and from simple to complex systems. Complexity science may integrate simple theory as its special case or an approximation at the lower order. Developing a new economic math is aimed to expanding our scope in economic thinking.

The fundamental debate between the exogenous and endogenous school in economics, or between the outside and inside view in philosophy, can be resolved by Prigogine's idea of "order through fluctuations" [66]. From the philosophical perspective of nonequilibrium physics, economies are open dissipative systems in nature. Structural evolution inside the system is strongly influenced by environmental changes outside the system. From mathematical perspective, all our new findings of color chaos and the birth-death process are observed through the HP filter. Its nonlinear smooth trend acts as a Copernicus reference system in analyzing macro and financial index. Its economic implication is replacing rational expectations by market trends that are shaped by reflexivity [67], or social interaction between global environment and market players. In our new option-pricing model based on collective behavior, the constant risk-free interest rate is replacing by the trend price in our model based by the birth-death process [68].

There is always a trade-off between mathematical simplicity and empirical relevance. We choose the birth-death process because it is the simplest model in population dynamics that is solvable by master equation. Although the birth-death process is more complex than the Brownian motion model of the representative agent, but we obtain more knowledge of social psychology. By the similar token, we learn more about changing economies from time-frequency analysis than that from static statistics. We take the experimental approach in choosing proper level of computational complexity. There are so many innovations in computational algorithms and data mining.

We found that theoretical physics including nonlinear dynamics, quantum mechanics and statistical mechanics have more explaining power in economic analysis for three reasons.

First, theoretical physics has a general and consistent framework in theory, while other numerical algorithms only have technical merits without deep theoretical foundation. Only physics theory provides a unified framework for physics, chemistry, biology, medicine, and now economics.

Second, neoclassical economics chooses the wrong framework from physics. Classical mechanics started from Newton's law in mechanics that is universal when particle speed was far below light speed. Hamiltonian formulation of classical mechanics is valid only for conservative system without energy dissipation. Neoclassical economics borrowed the Hamiltonian formulation to justify utility and profit maximization in a utopian world without energy dissipation and technology innovation. Non-equilibrium physics realized the fundamental differences between conservative system with time symmetry and dissipative system with time asymmetry. Living and social systems are dissipative systems in nature, which are evolving in time and space. Therefore, non-equilibrium forces, such as unpredictable uncertainty [69] and creative destruction [70], have more weight than equilibrium forces, such as cost competition and arbitrage activity, in driving economic movements. Newton mechanics did not exclude nonlinear interaction. Physicists discovered deterministic chaos from both conservative system and dissipative system. However, the requirement of convex set in microeconomics simply rules out of possibility of multiple equilibrium and chaos. We should reconstruct theoretical foundation of microeconomics, so that it is compatible with observed patterns in macro and finance.

Third, empirical evidence of economic complexity and economic chaos impose severe limits to economic reductionism and methodological individualism. Nonlinear interaction and multi-layer structure uncover the importance of evolutionary perspective and holistic view of human activity. From evolutionary perspective, nonlinearity and non-equilibrium mechanism is mathematical representation of irreversibility and history. Complex patterns from mathematical representation are hard to understand if we do not have relevant knowledge in economic structure and social history. In this sense, nonequilibrium and nonlinear approach may serve as a bridge to the Two Cultures [71]. We learn a great deal from wide-range dialogue with biologists, historians, sociologists, anthropologists, and psychologists, in addition to economists from different schools.

6 Conclusion

In philosophy, economists always believe that economic behavior should be more complex than those in physics. In practice, economic math is much simpler than math in physics. Typical example is the ideal gas model, which is simplest in statistical mechanics. Standard statistical mechanics started with an ensemble with many particles, whose distribution is determined by the underlying dynamics. In contrast, macro and finance theory in economics only have one particle in the representative agent model with fixed probability distribution [72]. Therefore, economic theory is more naïve than physics in mathematical representation. We need judge the economic philosophy by their deeds rather than words.

Historically, mathematical representation goes hand in hand with technology and social changes. Economic math also undergoes rapid development from simple to more advanced stage. Mathematical representation plays an important role in economic theory and empirical analysis. Economic math models are competing in three aspects: empirical explaining power, mathematical elegance, and philosophical implications.

From the historical lessons in science, it is empirical relevance of mathematical model that defines the meaning of mathematical beauty and philosophical status. For example, Einstein's relativity theory has better empirical foundation than Newton's mechanics. The success of relativity theory changed mathematical status of non-Euclidean geometry and philosophical foundation of absolute universe. In economics, natural experiments, such as the Great Depression and the 2008 Financial Crisis, shake the belief in the self-stabilizing market. People demand new thinking in economics. The nonlinear and non-equilibrium nature of the modern economy will be finally accepted by the public and economics community.

There are three forces that may accelerate changes in economic thinking.

First, big data and increasing computer power will stimulate upgrading in mathematical representation from linear, static models to nonlinear, non-stationary models in economic theory.

Second, tremendous progress of complexity science has made increasing application in science, engineering, and medicine. Their successes will spread to economics and finance. The closed atmosphere in mainstream economics cannot be last long.

Third, neoclassical economics is mainly the product of Anglo-Saxon culture based on individualism. The 2008 financial crisis and global warming marked the end of consumerism. Both developed countries and developing countries face severe challenges from environment, resource, and poverty. Economic development has to adapt to diversified environmental conditions. This social awareness will speed up paradigm changes in economic thinking.

Acknowledgements The author thanks Emiliano Ippoliti and participants of the Workshop on Math, Finance, and Philosophy at Rome on June 12–13, 2014. The author also thanks stimulating discussions with Yuji Aruka, Edward Fullbrook, James Galbraith, David Hendry, Justin Lin, Richard Nelson, Edmund Philps, Andreas Pyka, Hong Sheng, Zhengfu Shi, Joseph Stiglitz, Yinan Tang, Wolfgang Weidlich, and Elsner Wolfram.

References

1. Prigogine, I.: Order Out of Chaos: Man's New Dialogue with Nature, Bantam (1984)
2. Chen, P.: Economic Complexity and Equilibrium Illusion: Essays on Market Instability and Macro Vitality. Routledge, London (2010)
3. Mirowski, P.: More Heat than Light, Economics as Social Physics, Physics as Nature's Economics. Cambridge University Press, Cambridge (1989)
4. Fullbrook, Edward: A Guide to What's Wrong with Economics. Anthem Press, London (2004)

5. Rosewell, B.: Economists 'make assumptions', BBC NEWS Interview One of the Ten Leading Economists Who Wrote a Letter to the Queen on too Much Math Training in Economics. http://news.bbc.co.uk/today/hi/today/newsid_8202000/8202828.stm. Accessed 15 Aug 2009
6. Krugman, P.: How Did Economists Get It So Wrong? The New York Times, Sept 2 (2009)
7. Chen, P.: Equilibrium illusion, economic complexity, and evolutionary foundation of economic analysis, Evol. Inst. Econ. Rev. **5**(1), 81–127 (2008) (Ibid.2, Chapter 2)
8. Becker, G.: A note on restaurant pricing and other examples of social influences on price. J. Polit. Econ. **99**, 1106–1116 (1991)
9. Dessing, M.: Labor supply, the family and poverty: the s-shaped labor supply curve. J. Econ. Behav. Organ. **49**(4), 433–458 (2002)
10. Fama, E.F.: Efficient capital markets: a review of theory and empirical work. J. Financ. **25**(2), 384–433 (1970)
11. Sharp, W.F.: Capital asset prices: a theory of market equilibrium under conditions of risk. J. Financ. **19**(3), 425–442 (1964)
12. Mandelbrot, B.: The Variation of Certain Speculative Prices. J. Bus. **36**(4), 394–419 (1963)
13. Gabaix, X., Gopikrishnan, P., Plerou, V., Eugene Stanley, H.: Institutional investors and stock market volatility. Q. J. Econ. **121**(2): 461–504 (2006)
14. Jackwerth, J., Rubinstein, M.: Recovering probability distribution from option price. J. Financ. **51**, 1611–1631 (1996)
15. Haken, H.: Synergetics, An Introduction. Springer, Berlin (1977)
16. Weidlich, W.: Ising model of public opinion. Collect. Phenom. **1**, 51 (1972)
17. Chen, P.: Imitation, learning, and communication: central or polarized patterns in collective actions. In: Babloyantz, A. (ed.) Self-organization, Emerging Properties and Learning, pp. 279–286. Plenum, New York (1991) (Ibid.2, Chapter 9)
18. Samuelson, P.A.: Interactions between the multiplier analysis and the principle of acceleration. Rev. Econ. Stat. **21**, 75–78 (1939)
19. Frisch, R.: Propagation problems and impulse problems in dynamic economics. In: Economic Essays in Honour of Gustav Cassel, George Allen & Unwin, London (1933)
20. Chen, P.: Empirical and theoretical evidence of monetary chaos. Syst. Dyn. Rev. **4**, 81–108 (1988) (Ibid.2, Chapter 4)
21. Chen, P.: Microfoundations of macroeconomic fluctuations and the laws of probability theory: the principle of large numbers vs. rational expectations arbitrage. J. Econ. Behav. Organ **49**, 327–344 (2002) (Ibid.2, Chapter 13)
22. Samuelson, P.A.: Interactions between the multiplier analysis and the principle of acceleration. Rev. Econ. Stat. **21**, 75–78 (1939)
23. Box, G.E.P., Jenkins, G.M.: Time Series Analysis, Forecasting and Control. Holden-Day, San Francisco (1970)
24. May, R.M.: Simple mathematical models with very complicated dynamics. Nature **261**(5560), 459–467 (1976)
25. Day, R.H.: Irregular growth cycles. Am. Econ. Rev. **72**, 404–414 (1982)
26. Henon, M.: A two dimensional mapping with a strange attractor. Commun. Math. Phys. **50**, 69–77 (1976)
27. Benhabib, J.: Adaptive monetary policy and rational expectations. J. Econ. Theory **23**, 261–266 (1980)
28. Lorenz, E.N.: Deterministic nonperiodic flow. J. Atmos. Sci. **20**, 130–141 (1963)
29. Mackey, M.C., Glass, L.: Oscillations and chaos in physiological control systems. Science **197**, 287–289 (1977)
30. Chen, P.: Empirical and theoretical evidence of monetary chaos. Syst. Dyn. Rev. **4**, 81–108 (1988) (Ibid.2, Chapter 4)
31. Rössler, O.E.: An equation for continuous chaos. Phys. Lett. A **57**, 397–398 (1976)
32. Samuelson, P.A.: Interactions between the multiplier analysis and the principle of acceleration. Rev. Econ. Stat. **21**, 75–78 (1939)

33. Chen, P.: Empirical and theoretical evidence of monetary chaos. Syst. Dyn. Rev. **4**, 81–108 (1988) (Ibid.2, Chapter 4)
34. Nelson, C.R., Plosser, C.I.: Trends and random walks in macroeconomic time series, some evidence and implications. J. Monet. Econ. **10**, 139–162 (1982)
35. Chen, P.: Metabolic growth theory: market-share competition, learning uncertainty, and technology wavelets. J. Evol. Econ. **24**(2), 239–262 (2014)
36. Schumpeter, J.A.: The Theory of Economic Development. Harvard University Press, Cambridge (1934)
37. Smith, A.: The Wealth of Nations, Liberty Classics, Indianapolis, Book I, Chapter III, Division of Labor is Limited by Market Extent (1776, 1981)
38. Chen, P.: Microfoundations of macroeconomic fluctuations and the laws of probability theory: the principle of large numbers vs. rational expectations arbitrage. J. Econ. Behav. Organ. **49**, 327–344 (2002) (Ibid.2, Chapter 13)
39. Malkiel, B.G.: A Random Walk Down Wall Street: Norton (2003)
40. Black, F., Scholes, M.: The pricing of options and corporate liabilities. J. Polit. Econ. **81**, 637–654 (1973)
41. Reichl, L.E.: A Modern Course in Statistical Physics, 2nd edn. Wiley, New York (1998)
42. Chen, P.: Evolutionary Economic Dynamics: Persistent Business Cycles, Disruptive Technology, and the Trade-Off between Stability and Complexity. In: Dopfer, K. (ed.) The Evolutionary Foundations of Economics, Chapter 15, pp. 472–505. Cambridge University Press, Cambridge (2005) (Ibid.2, Chapter 3)
43. Zeng, W., Chen, P.: Volatility smile, relative deviation and trading strategies: a general diffusion model for stock returns based on nonlinear birth-death process. China Econ. Q. **7**(4), 1415–1436 (2008)
44. Tang, Y., Chen, P.: Option pricing with interactive trend and volatility. Financ. Rev. **2**, 1–11 (2010)
45. Frisch, R.: Propagation problems and impulse problems in dynamic economics. In: Economic Essays in Honour of Gustav Cassel, George Allen & Unwin, London (1933)
46. Uhlenbeck, G.E., Ornstein, L.S.: On the theory of brownian motion. Phys. Rev. **36**(3), 823–841 (1930)
47. Wang, M.C., Unlenbeck, G.E.: On the theory of the Brownian Motion II. Rev. Mod. Phys. **17** (2&3), 323–342 (1945)
48. Chen, P.: The Frisch Model of Business Cycles: A Failed Promise and New Alternatives, IC^2 Working Paper. University of Texas at Austin (1998) (Ibid.2, Chapter 12)
49. Hodrick, R.J., Prescott, E.C.: Post-War US. Business Cycles: An Empirical Investigation, Discussion Paper No. 451, Carnegie-Mellon University (1981) (J. Money Credit Bank. **29**(1), 1–16 (1997))
50. Von Neumann, J., Kent, R.H., Bellinson, H.R., Hart, B.I.: The mean square successive difference. Ann. Math. Stat. **12**(2), 153–162 (1941)
51. Chen, P.: Economic Complexity and Equilibrium Illusion: Essays on Market Instability and Macro Vitality, Chapter 2, p. 19. London: Routledge (2010) (Ibid.2, Chapter 2)
52. Chen, P.: A random walk or color chaos on the stock market?—Time-frequency analysis of S&P indexes. Stud. Nonlinear Dyn. Econ. **1**(2), 87–103 (1996) (Ibid.2, Chapter 6)
53. Schumpeter, J.A.: Business Cycles, A Theoretical, Historical, and Statistical Analysis of the Capitalist Process. McGraw-Hill, New York (1939)
54. Schrödinger, Enwin: What is Life?. Cambridge University Press, Cambridge (1948)
55. Chen, P.: Microfoundations of macroeconomic fluctuations and the laws of probability theory: the principle of large numbers vs. rational expectations arbitrage. J. Econ. Behav. Organ. **49**, 327–344 (2002) (Ibid.2, Chapter 13)
56. Chen, P.: From an efficient market to a viable market: new thinking on reforming the international financial market. In: Garnaut, R., Song, L., Woo, W.T. (eds.). China's New Place in a World in Crisis: Economic, Geopolitical and the Environmental Dimensions, Chapter 3, pp. 33–58. Australian National University E-Press and The Brookings Institution Press, Canberra, July 14 (2009) (Ibid.2, Chapter 16)

57. Johnson, S.: The quiet coup. Atlantic **303**(4), 46–56 (2009)
58. Tang, Y., Chen, P.: Transition probability, dynamic regimes, and the critical point of financial crisis. Phys. A (2015)
59. Tang, Y., Chen, P.: Time varying moments, regime switch, and crisis warning: the birth-death process with changing transition probability. Phys. A **404**, 56–64 (2014)
60. Prigogine, I.: From Being to Becoming: Time and Complexity in the Physical Sciences. Freeman, San Francisco (1980)
61. Kuhn, T.: The Structure of Scientific Revolutions, Chicago (1962)
62. Foley, D.K.: Adam's Fallacy: A Guide to Economic Theology. Harvard University Press (2008)
63. Chen, P.: Metabolic growth theory: market-share competition, learning uncertainty, and technology wavelets. J. Evol. Econ. **24**(2), 239–262 (2014)
64. Friedman, M.: Essays in Positive Economics. University of Chicago Press, Chicago (1953)
65. Hendry, D.F.: Econometrics: Alchemy or Science?, 2nd edn. Oxford University Press, Oxford (2001)
66. Nicolis, G., Prigogine, I.: Self-organization in Nonequilibrium Systems: From Dissipative Structures to Order through Fluctuations. Wiley, New York (1977)
67. Soros, G.: Reflexivity in financial markets. In: The New Paradigm for Financial Markets: The Credit Crisis of 2008 and What it Means, 1st edn. PublicAffairs (2008)
68. Tang, Y., Chen, P.: Option pricing with interactive trend and volatility. Financ. Rev. **2**, 1–11 (2010) (in Chinese)
69. Knight, F.H.: Risk, Uncertainty and Profit, Sentry Press, New York (1921)
70. Schumpeter, J.A.: Capitalism, Socialism and Democracy, 3rd edn. Harper, New York (1950)
71. Chen, P.: Nonequilibrium and nonlinearity: a bridge between the two cultures. In: Scott, G.P. (ed.) Time, Rhythms, and Chaos in the New Dialogue with Nature, Chapter 4, pp. 67–85. Iowa State University Press, Ames (1980)
72. Black, F., Scholes, M.: The pricing of options and corporate liabilities. J. Polit. Econ. **81**, 637–654 (1973)

Behind the Price: On the Role of Agent's Reflexivity in Financial Market Microstructure

Paolo Barucca and Fabrizio Lillo

Abstract In this chapter we review some recent results on the dynamics of price formation in financial markets and its relations with the efficient market hypothesis. Specifically, we present the limit order book mechanism for markets and we introduce the concepts of market impact and order flow, presenting their recently discovered empirical properties and discussing some possible interpretation in terms of agent's strategies. Our analysis confirms that quantitative analysis of data is crucial to validate qualitative hypothesis on investors' behavior in the regulated environment of order placement and to connect these micro-structural behaviors to the properties of the collective dynamics of the system as a whole, such for instance market efficiency. Finally we discuss the relation between some of the described properties and the theory of reflexivity proposing that in the process of price formation positive and negative feedback loops between the cognitive and manipulative function of agents are present.

1 Introduction

Understanding price movements, both their origin and their properties, is one of the most important challenges in economics. In Finance this problem has a long history which has seen two antagonist schools of thought confronting in the first half of the twentieth century. On one side there were the fundamentalists who posited that the price of an asset is the discounted value of its "intrinsic" or "fundamental" value and

---

Authors acknowledge partial support by the grant SNS13LILLB "Systemic risk in financial markets across time scales".

P. Barucca (✉) · F. Lillo
Department of Banking and Finance University of Zurich, Rämistrasse 71, 8006 Zürich, Switzerland
e-mail: paolo.barucca@bf.uzh.ch

F. Lillo
QUANTLab, via Pietrasantina 123, 56122 Pisa, Italy

E. Ippoliti and P. Chen (eds.), *Methods and Finance*,
Studies in Applied Philosophy, Epistemology and Rational Ethics 34,
DOI 10.1007/978-3-319-49872-0_3

equals the discounted cash flow which that security gives title to. On the other side there were the econometricians, who, applying statistical analyses to empirical time series of prices discovered that stock prices develop patterns that look like those of a random walk. The latter is the time series that can be obtained, for example, by tossing a coin and moving up (down) the price by one unit when the outcome is head (tail). The erratic and random behavior of prices seemed to clash with the fundamentalist view and seemed to give support to those believing that stock market is essentially a casino. By using the words of LeRoy "If stock prices were patternless, was there any point to fundamental analysis?". It is well known that the solution to this problem was given by the seminal 1965 paper by Paul Samuelson "Proof that Properly Anticipated Prices Fluctuate Randomly" (even if Bachelier 1900 and Cowles 1933 arrived to somewhat similar conclusions). Samuelson showed that in an informationally efficient market, i.e. a market which fully incorporates the expectations and information of all market participants, price changes must be unforecastable. This is the celebrated Efficient Market Hypothesis (EMH), a cornerstone of modern Finance, which reconciles the fundamentalist and econometrician view.

However the details of how information is impounded into price are still a matter of debate, as well as the question of whether markets are truly efficient. Market microstructure is devoted to the empirical and theoretical study of how information (private or public) is incorporated into price and how price is formed through the action of many interacting agents. Given the importance of the problem (even outside Finance, think for example to the new markets for advertisement on search engines in the Internet), the increasing availability of high resolution data, and the significant changes observed in the organization of markets, market microstructure has experienced a large development in the last fifteen years. In this paper we review the problem of price formation from a microstructure point of view focusing on how strategies are slowly translated into orders in the market. We do not aim to be exhaustive, but to highlight the main elements which have recently emerged in the field, trying to avoid as much as possible technicalities and the use of mathematical formalism. In the last part of the paper we will discuss about possible analogies and relations between the newly discovered properties of price formation and the theory of social reflexivity, proposed, among others, by George Soros.

The contribution is divided in four sections. In the first section we present the efficient market hypothesis, the concept of market impact, and we describe a widespread market mechanism, namely the limit order book. In the second section we present the order flow and describe one important properties, its long memory, which allows us to understand the strategy of order splitting followed by large investors. In this section we also discuss how the long memory of order flow can be reconciled with efficient market hypothesis. In the third section we review Soros' theory of social reflexivity and finally in the last section we outline our interpretation of empirical results in the context of social reflexivity.

2 Market Impact and the Limit Order Book

The market is where supply and demand meet, that is where buyers and sellers exchange goods. The efficient-market hypothesis (EMH) states [1, 2] that stock market efficiency implies existing share prices to always incorporate and reflect all relevant information. According to EMH, any information in the hand of an investor on the future value of an assets should thus be instantaneously incorporated into the price. This occurs through the choice of the strategy the investor chooses to trade the asset. The trading strategy can be seen as the medium that translate information into price changes. Loosely speaking *market impact* refers to the correlation between an incoming order or strategy (to buy or to sell) and the subsequent price change. The strategy can be coded in the series of *orders* to buy or to sell, which are sent to the market. The *order flow* is the aggregation of the orders of all the agents (or a subset of them).

However, the causal relation between trading strategies (or order flow) and price changes is far from trivial and at least three explanations for this correlation (i.e. the existence of market impact) can be given

- Agents successfully forecast short term price movements and trade accordingly. This does result in measurable correlation between trades and price changes, even if the trades by themselves have absolutely no effect on prices at all. If an agent correctly forecasts price movements and if the price is about to rise, the agent is likely to buy in anticipation of it. According to this explanation, trades with no information content have no price impact.
- The impact of trades reveals some private information. The arrival of new private information causes trades, which cause other agents to update their valuations, leading to a price change. But if trades are anonymous and there is no easy way to distinguish informed traders from non informed traders, then all trades must impact the price since other agents believe that at least of fraction of these trades contains some private information, but cannot decide which ones.
- Impact is a purely statistical effect. Imagine for example a completely random order flow process, that leads to a certain order book dynamics. Conditional to an extra buy order, the price will on average move up if everything else is kept constant. Fluctuations in supply and demand may be completely random, unrelated to information, but a well defined notion of price impact still emerges. In this case impact is a completely mechanical—or better, statistical—phenomenon.

The distinction between these three scenarios is not fully understood. In order to discriminate among these alternatives, it is useful to specialize into a concrete case, since markets have different structures and rules. In other words, one way to address this problem in a quantitative fashion is through dynamical models of market microstructure, which are based on a specific, 'internalist', knowledge of what kind of orders can be placed by investors and what is their dynamics. Here we will focus on the *limit order book*, a very common mechanism adopted in most electronic markets.

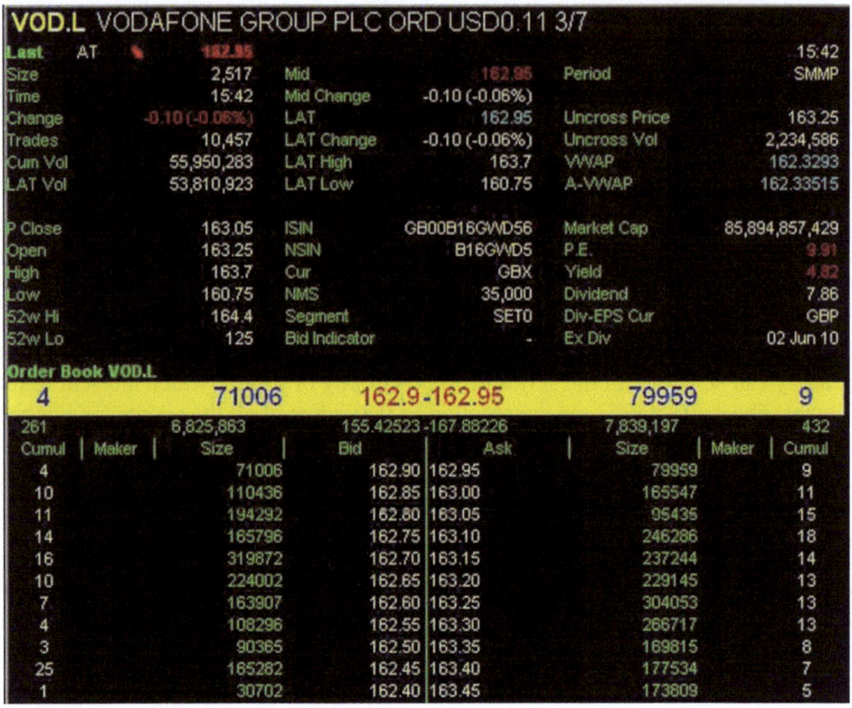

Fig. 1 The typical structure of an order book [3]

In a limit order book market, an agent (or an intermediary acting on her behalf) can place two types of orders,[1] namely limit and market orders:

- Limit orders are orders to buy or sell a given volume of shares at a specified price or better.
- Market orders are orders placed to buy or sell an investment immediately at the best available current price.

The best buy and sell order on the market are called ask and bid respectively and their difference is the bid-ask spread. Buy (sell) limit orders with a price lower (higher) than the ask (bid) do not initiate a transaction and are stored in queue, visible to other market participants, waiting for the price moving in their direction. An agent can decide to cancel a limit order at any time. Figure 1 shows a snapshot of a real order book.

Orders are not typically placed directly by individual investors, mutual funds or hedge funds: all these economic agents need to ask intermediaries to place orders of given volumes of various stocks on their behalf. Other subjects that can place orders in the market are market makers who act as intermediaries for the market as a whole: they continuously place orders in the order book so to provide 'liquidity'

[1]Even though various financial markets may have more kinds of slightly different orders, these are the two main types-.

to the market, that is to give investors the chance to buy and sell anytime at their will. Market makers take risks by buying and selling countercurrent but make profit through the bid-ask spread.

Market impact arises for two reasons. The first is purely mechanical. A market order with a volume larger than or equal the volume at the opposite best (the bid for a sell market order and the ask for a buy market order) will move mechanically the price, increasing the spread and therefore the mid price.[2] Notice that even an uninformed trade moves mechanically the price and there is no relation between information of the trade and price movement (see the third explanation in the bulleted list above). This is even reinforced by the fact that electronic markets are typically anonymous and there is no way of knowing the identity of the counterpart.

The second possible origin of market impact is due to the reaction of the rest of the market to the trade. Even if the market order does not move mechanically the price, the other agents might react to the observation of the trade, revising the price of their limit orders and thus moving the price. As we will see below, the induced market impact plays an important role. In general, the more liquid is the market, i.e. the more limit orders are present in the order book then the lower is the market impact and the less a single investor can induce large price changes.

A number of empirical studies has [4–6] established that market impact, both mechanical and induced, is statistically different from zero (despite being quite noisy). This means that buy (sell) market orders move on average the price upward (downward). Generically it is found that the larger the volume of the trade the larger on average the price change and the dependence between these two quantities is strongly concave. In simple words this means that if the volume of a market order is doubled, the price impact is significantly less than doubled. The fact that even uninformed trades can change the price (again, also because it is almost impossible that the market understands quickly if the trader is informed or not) raises several interesting questions on the relation between information and price changes and therefore on the Efficient Market Hypothesis. Anticipating the discussion we will do later in this paper, it suggests the presence of positive feedback loops, where an uninformed trade transiently impacts price, but the other market participants, being unable to discern if the trade is informed or not, revise their valuation of the asset and trade concurrently, creating more price change in the direction of the random initiator trade. This amplification mechanism resembles the reflexivity hypothesis (see below for more details).

Note that all the above analysis is static, i.e. it refers to the contemporaneous relation between trades and price changes. An important aspect to elucidate better the process with which markets understand if a trade is informed or not and how the information (or the lack thereof) is incorporated into price requires an inter-temporal analysis, investigating how the orders arrive to the market sequentially in time and

[2]Note also that if market order volume is smaller than or equal to the volume at the opposite best, the order is executed at the best price; on the other hand if its volume exceeds the volume of the opposite best in the order book, then it penetrates the book and reaches the second best price or more. In this case the price is a weighted average over the various limit orders that are needed to execute the market order.

how they impact the price. Next section discusses some recent findings in this direction.

3 The Long Memory of the Market

In the last ten years there have been major efforts to understand investors behaviors through detailed modeling and data-analysis. A significant amount of the literature has been devoted to analyze data on order books of real financial markets, [4, 7]. In particular we review here the empirical properties of order flow and its consequences for the modeling and understanding of market impact.

Understanding the inter-temporal properties of the order flow is quite challenging. This is due to the intrinsic multi dimensional structure (each type of order is a different variable) whose properties depend on the current state of the book, which in turn is the result of the past order flow. The order flow and limit order book modeling is still an open problem and new models and empirical results are continuously proposed.

Here we focus on a small but important subset of the order flow. Specifically we shall consider the flow of market orders (hence the orders triggering a transaction) and we discard the information on the volume of the order, focusing only on its sign. The sign $\epsilon(t)$ of the t-th trade is equal to $+1$ for a buy order and -1 for a sell order. Thus the binary time series of $+1$ and -1 is the simplest encoding of the dynamics of supply and demand for a given asset. One important question is what are the time autocorrelation properties of such time series. Technically speaking the autocorrelation function of the signs $\epsilon(t)$ is the expectation (or the mean) of the product of two transaction signs separated by τ transactions. For a totally random sequence (for example the one obtained by tossing a coin), this function is equal to zero for all τs because the occurrence of a head now does not give any information about the likelihood that a head or a tail will occur τ tosses in the future. The autocorrelation function is therefore related to the predictability of trade signs.

A series of papers in the last ten years have shown [8–10] that in the vast majority of financial markets the autocorrelation function of the order flow is a slowly decaying function of the lag τ, well described by a form $\tau^{-\gamma}$, where γ is an exponent empirically found to be between 0 and 1. These kind of processes are called long memory processes because it can be shown that they lack a typical time scale beyond which the process appears to be random. In other words the present state of the system is determined by the whole past history of the process. The exponent γ measures the memory of the process, which is slowly decaying for large τ. It is possible to quantify this slowness from the empirical order flow auto-covariance: the smaller the exponents, the slower the decay. This slowness is commonly interpreted and quantified through the Hurst exponent H. In the case of random diffusion, where no memory is present and order signs are drawn randomly with equal probability at each time-step, the Hurst exponent is $1/2$. For long-memory processes H is larger than $1/2$, as it is for the order flow so that the it can be regarded as a super-diffusive process, where fluctuations grows with time through an exponent larger than $1/2$.

The result is very robust and it has been checked independently by different research groups and with different statistical methods. The observed values of H vary across markets but they alway remain larger than $1/2$. Long-memory processes have been observed also in the dynamic behavior of other financial variables: volatility of prices and trading volume have been recognized as long-memory processes [11].

The observation of long memory of order flow raises two interesting questions: (i) what is the behavioral origin for the sign persistence in the order flow and (ii) how is it possible to reconcile the predictable order flows with market efficiency (i.e. unpredictability of price changes).

The present explanations for long-memory fall into two classes: the first is that the long memory of the order flow holds for each investor and it links the persistence in the order flow with the presence of large meta-orders that are splitted in subsequent small orders of the same sign; the second class of explanations calls into question collective behaviors of investors imitating each other [12]. Evidence gathered so far seems to favor the first class of explanations, in particular data show that large investors do split and execute their orders. Furthermore predictions concerning the relation between trade volumes, market impact and order flow are in agreement with the first type of explanations [13].

The presence of meta-orders is indeed a simple and clear explanation for long-memory in the order book. Let us consider an investor that has to execute a large trade and who does not want to influence the price abruptly. Instead of placing directly a big market order that would penetrate the order book and reach limit order at strongly unfavorable prices, the investor prefers to split the meta-order and both gain the chance not to influence the market and to get better prices if on average other investors are not following the same strategy.

The other important question is how the long memory of order flow is consistent with EMH. Since a buy (sell) trade moves the price upward (downward) a persistent order flow sign time series would induce a persistent price change time series. This means that by observing that recent past price changes were, for example, typically positive one could predict that in the near future the price will move up. This (hypothetical) predictability would allow to make easy profits and is inconsistent with EMH and with empirical observation.

A possible solution of this apparent paradox is the asymmetric liquidity mechanism [13]. According to it the price impact of a trade depends on the predictability of its sign. For example, if past trades were typically buys, implying that the next trade is more likely a buy than a sell, the asymmetric liquidity postulates that if the more predictable event (the buy) actually occurs, it will impact the price less than if the less likely event (the sell) occurs. In other words the price change is not fixed, but it is history dependent and adapts itself to the level of predictability of its source (the order flow).

The asymmetric liquidity mechanism has been verified empirically [13] and its microscopic origin has been explored and elucidated in [14]. In this last paper it has been shown that agents executing incrementally a large order by splitting it in a large number of trades adjust the size of these trades in order to minimize their impact on price. This adjustment becomes stronger and stronger as the execution proceeds. In other words investors decide their strategy exactly because they are conscious of their impact on price. Finally, it has been shown that this mechanism is able to reconcile the long memory of the order flow with the uncorrelated price change time series.

In conclusion, the splitting of metaorders and its detection from the rest of the market is critical to understand the dynamics of price in financial markets. From a strategic point of view, the origin of splitting has been explained and motivated theoretically in the seminal work of Kyle [15]: the optimal strategy for an investor with private information about the future price of an asset is to trade incrementally through time. This strategy allows earlier executions to be made at better prices and minimizes execution cost. Moreover this strategy minimizes the information leakage, i.e. it allows the trader to hide her intention (and information). This strategy can be seen as a different form of reflexivity. In fact the purpose of splitting is to modify, with the action of trading and the impact it generates, the beliefs of the other market participants. Differently from the impact case described in the previous section, here this form of interaction between action and beliefs creates a negative feedback. In the next two sections we discuss in more detail the relations between these empirical facts and the theory of reflexivity.

4 Theory of Social Reflexivity

In social systems agents do not merely observe but also actively participate in the system themselves, this is the simple observation leading to reflexivity theory in social sciences. Soros first exposed his theory in his book *The Alchemy of Finance* in 1987 where he builds his conceptual framework on two principles. The first one is the principle of fallibility: in social systems the participants' views and consequently their perspectives are always biased, inconsistent, or both.

Human beings are utterly familiar with the principle of fallibility that affects all aspects of human life, from perception to action. Fallibility is strictly connected to the world *complexity*; social facts are so complex that cannot be perceived and understood completely and thus they require a simplification that introduces biases and inconsistencies in our understanding. But fallibility is not just about perception.

If we interpret fallibility as *the inability of a human being to elaborate an optimal response in a given social situation* then we can distinguish different causes of fallibility: (i) A subjective cognition fallibility: we are unable to act in the best possible way because we perceive our situation in a subjective and incomplete fashion; (ii) a general cognition fallibility: we perceive the situation of all the other social agents in a subjective and incomplete fashion; (iii) a manipulative fallibility: even assuming a perfect perception of the system situation, we can act in a wrong way.

The second is the principle of reflexivity: the imperfect views of participants can influence the situation to which they relate through their actions. In particular Soros gives [16] this specific example: "if investors believe that markets are efficient then that belief will change the way they invest, which in turn will change the nature of the markets in which they are participating (though not necessarily making them more efficient)."

The role of reflexivity can be described with a circle of social actions, all affected by fallibility: the social agent perceives a given situation, formulates an interpretation, decides a strategy, and finally acts but by acting she changes the situation so that the act might no longer have the same effect that was hypothesized at first.

This circle can generate two kinds of feedback loops, negative or positive. Negative feedback loops of participants' actions can stabilize the system, in the sense that the situation of the system becomes more and more similar to the participants' perceptions thanks to their actions, thus helping to reach an equilibrium.

Conversely positive feedback loops destabilize the system, since participants' perceptions and the real situation of the system differ bringing the system *far from equilibrium* towards an eventual sudden change, e.g. the case of *bubbles* and *crashes* in financial markets.

Indeed financial markets are excellent and profitable laboratories to test reflexivity theory, and cases of positive feedback loops are investigated in [4]. In the next section we will discuss some of the previously described properties of the price formation mechanism, market impact in the order book and order splitting, as examples of reflexivity where positive and negative feedback loops might play a major role.

5 Discussion on Reflexivity and Price Formation

In the first two sections we have given a general introduction and a specific description of the intriguing phenomenology of price formation in financial markets, limit order books, market impact, and order flow. We have discussed the possible origin of market impact, and the still open issue of its cause. We have also presented the properties of order flow, i.e. the dynamics of supply and demand arriving into the market. We have presented the long-memory property of order flow and explained it as a consequence of the splitting of meta-orders by investors. Furthermore we have shown how the dependence of market impact from trade predictability can explain the coexistence of long-memory in the order flow and the fast decay of autocorrelation of price changes. Even if the description is by necessity short and synthetic, we hope that we convinced the readers that the process of price formation is interesting and still not fully understood. It is obviously at the heart of many economic (and not necessarily only financial) phenomena and has a significant number of practical consequences.

In the text we have also proposed some analogies between the some elements of price formation and the theory of social reflexivity (reviewed in Sect. 3). The decisions of investors in financial markets depend on their beliefs on the traded assets and clearly price plays an important role of signal in this cognitive activity of agents. However price is affected, via market impact, by the decision of the investors and this creates the simultaneous presence of the manipulative and the cognitive function of humans, a key condition, according to Soros, for social reflexivity [16, 17]. We therefore believe that (financial) market microstructure is a perfect playground for studying reflexivity and understanding feedback loops between these two functions.

More in detail, in this chapter we have sketched two possible mechanisms for reflexivity in price formation. The first is at the core of microstructure, since it concerns the origin of market impact. In fact, price moves in response to informed trades, but, at least on the short term, it moves also mechanically as a consequence of trading orders. Since other market participants cannot discern informed trades, also uninformed trades are able to move the price. This manipulative function modifies the cognitive activity of the other market participants, who revise their valuation of the asset as a consequence of the impact of a (possibly uninformed) trade. This process creates a positive feedback loop where small price fluctuations, even when generated by uninformed trades, can be amplified by the activity of reflexive agents, and this process can create price bubbles or short term large price fluctuations.

The second mechanism for reflexivity is related to the activity of order splitting. We have seen that, in consequence of the small liquidity present in financial markets and of the existence of market impact, investors who want to trade large quantities split their orders in small pieces and trade them incrementally over periods of time ranging from few minutes to several days. Other agents continuously monitor the flow of orders arriving to the market with the objective of identifying such splitting activities. This is because (i) the splitting activity of a large investor can signal her information on the future value of the price and (ii) knowing that a large investor is unloading an order gives the opportunity of front loading the investor, i.e. trading quickly with the hope to be the counterpart of the large investor in the future (and realizing a profit). Also in this case there are several interactions between the cognitive and manipulative functions of the agents. The large investor has a belief on the future price of an asset and through her trading moves the price in that direction. The other agents monitor the price and the order flow, learning the presence of a large investor and her belief on the price. Through their trading activity they modify the price pattern which in turn can modify the beliefs of other agents (and even of the splitting strategy of the large trader).

The two mechanisms presented here are clearly not exhaustive of the possible role of reflexivity in price formation and market microstructure. One of the lessons that can be learnt from this type of analysis is that the knowledge and modeling of the detailed mechanism through which agents interact is critical to understand some of the most important processes in economics and interaction of social agents. The second is that quantitative analysis of data is fundamental to validate qualitative hypothesis on investors' behavior in the market, to connect these micro-structural behaviors to market efficiency, and to formulate new hypotheses about the founding features of social systems.

References

1. Fama, Eugene F.: Market efficiency, long-term returns, and behavioral finance. J. Financ. Econ. **49**(3), 283–306 (1998)
2. Fama, E.F.: Efficient capital markets: a review of theory and empirical work. J. Financ. **25**(2), 383–417 (1970)
3. London Stock Exchange Group website. http://www.lseg.com
4. Farmer, J.D., Gerig, A., Lillo, F., Waelbroeck, H.: How efficiency shapes market impact. Quant. Finan. **13**(11), 1743–1758 (2013)
5. Zarinelli, E., Treccani, M., Farmer, J.D., Lillo, F.: Beyond the square root: evidence for logarithmic dependence of market impact on size and participation rate. ArXiv preprint arXiv:1412.2152 (2014)
6. Toth, B., Lemperiere, Y., Deremble, C., De Lataillade, J., Kockelkoren, J., Bouchaud, J.-P.: Anomalous price impact and the critical nature of liquidity in financial markets. Phys. Rev. X **1**(2), 021006 (2011)
7. Bouchaud, J.-P., Gefen, Y., Potters, M., Wyart, M.: Fluctuations and response in financial markets: the subtle nature of 'random' price changes. Quant. Finan. **4**(2), 176–190 (2004)
8. Farmer, J.D., Lillo, F.: On the origin of power-law tails in price fluctuations. Quant. Finan. **4**(1), 7–11 (2004)
9. Lillo, F., Mike, S., Farmer, J.D.: Theory for long memory in supply and demand. Phys. Rev. E **71**(6), 066122 (2005)
10. Bouchaud, J.-P., Farmer, J.D., Lillo, F.: How markets slowly digest changes in supply and demand. Fabrizio, How Markets Slowly Digest Changes in Supply and Demand (2008)
11. Cont, R.: Empirical properties of asset returns: stylized facts and statistical issues. 223–236 (2001)
12. LeBaron, B., Yamamoto, R.: Long-memory in an order-driven market. Phys. A: Stat. Mech. Appl. **383**(1), 85–89 (2007)
13. Lillo, F., Farmer, J.D.: The long memory of the efficient market. Stud. Nonlinear Dyn. Econometrics **8**(3) (2004)
14. Taranto, D.E., Bormetti, G., Lillo, F.: Limit order books. J. Stat. Mech. Theor. Exp. **2014**(6), P06002 (2014)
15. Kyle, A.S.: Continuous auctions and insider trading. Econometrica 1315–1335, 1985
16. Soros, George: Fallibility, reflexivity, and the human uncertainty principle. J. Econ. Methodol. **20**(4), 309–329 (2013)
17. Soros, G.: The alchemy of finance. Wiley (2003)
18. Farmer, J.D., Patelli, P., Zovko, I.I.: The predictive power of zero intelligence in financial markets. Proc. Nat. Acad. Sci. U.S.A. **102**(6), 2254–2259 (2005)

On the Description of Financial Markets: A Physicist's Viewpoint

Francesco Sylos Labini and Sergio Caprara

Abstract The main concepts at the foundation of the so-called neoclassical description of financial markets are here scrutinized from a physicist's viewpoint [1, 2]. Indeed, in the last two decades, physicists have devoted an increasing activity to the scrutiny of ideas and concepts that are at the basis of that branch of economic theory customarily called neoclassical—at the foundation of the neoliberal doctrine —that appears to be culturally and politically dominating in the present hard times. This activity appeared as surprising in the early days of its rise and development, since the objects studied by physicists are usually atoms, molecules, planets or galaxies, that look quite different from the objects studied by social sciences, the queen of which is economics. Human beings, contrary to elementary particles or stars, are endowed with free will and, more important, the laws that rule the ways in which an individual makes her/his own choices and by which different individuals establish relations among them, developing a social behaviour, are unknown to us. Rather, it seems legitimate to doubt that such laws are well defined.

In fact, we know the fundamental laws that rule, for instance, the interactions between electrical charges, or between the planets and the Sun: such laws, like, e.g., gravity, are universal and are the same at different points in space and at different times. For sure, we cannot say the same about the laws ruling economy: it is enough to jump back in time a hundred years, or to consider the present situation in underdeveloped countries, to immediately realize that the laws of economy, that we do not know to the same extent as we can write the equation describing the gravitational force that exists between the Earth and the Sun anyway, change in time and space, according to the historical, social, and legislative conditions of different

F. Sylos Labini
Enrico Fermi Center and Institute of Complex System, CNR, Via dei Taurini 19, 00185 Rome, Italy

S. Caprara (✉)
Physics Department, University of Rome "Sapienza", Piazzale Aldo Moro 2, 00185 Rome, Italy
e-mail: caprara@roma1.infn.it

E. Ippoliti and P. Chen (eds.), *Methods and Finance*,
Studies in Applied Philosophy, Epistemology and Rational Ethics 34,
DOI 10.1007/978-3-319-49872-0_4

countries. Indeed, time in a physical theory lacks a historical perspective, whereas time in economics is, or should be, historical time. In the end, the cost of an apple nowadays has nothing to do with the cost of an apple during World War II, etc.

Furthermore, the differences are not limited to this aspect. Precisely by virtue of the universality of natural laws, in physics we can repeat an experiment to ascertain the causal relations within a given series of processes. Moreover, thanks to our knowledge of the natural laws that rule the dynamics of some phenomena, we can make predictions, to be tested against experiments carried out under controlled conditions, so as to rule out or minimize the effect of external factors (not included in the theory). The results of these experiments must be repeatable, given the same conditions, in every other place, at any other time. Validation of a theory through reproducible experiments stands, indeed, at the basis of the scientific method [3].

Given this situation, one may be led to conclude that economics is a discipline far different from a natural science, like physics, and therefore, exactly the same way an economist does not possess technical and conceptual instruments apt to contribute to a physical problem in a significant way, a physicist undertaking problems so different, in some sense more circumscribed, and methodologically more defined, has nothing interesting to say about the dynamics of economy, because incomparably more complex. This conclusion seems straightforward, in virtue of the fact that the objects of study in physics are inanimate, the laws of nature are universal and the experiments are reproducible.

Notwithstanding the huge differences between the two disciplines, there are some concepts and methods that have been developed in physics—and more generally in natural sciences—in the last century, which may and should be a patrimony common to social scientists and economists too. We aim indeed at providing a brief guideline apt to highlight the impact of these concepts on neo-classical economics.

Among the concepts developed by physicists that may have quite an impact on social sciences we recall: deterministic chaos, that is at the basis of the behaviour of systems with nonlinear interactions, such as the Earth-Moon-Sun system, or the atmosphere, in which small perturbations may cause enormous and surprising effects; scale-invariant systems, like fractals, for which the customary notions of statistics, like mean and variance, must be abandoned to focus on scaling exponents; systems characterized by large fluctuations, in which extreme events are far more probable than usual; systems out of equilibrium, that are intrinsically unstable and for which the stable (steady) equilibrium state becomes irrelevant; systems that self-organize in "critical" states, characterized by scale-invariant dynamics, and subject to sudden and dramatic changes; systems in which adaptation and diversification are the crucial dynamical elements. These concepts, and many others, have been introduced, and are by now customary, in physics, meteorology, geology, biology [4].

On the other hand, it does not seem that these concepts have been appreciated by neoclassical economists: indeed, the core of this doctrine is the concept of equilibrium. That is, it is assumed that the economic system is very close to a situation of stable equilibrium and thus any fluctuation perturbing its state can be rapidly absorbed by the system itself, as it occurs when a gas in an isolated container is slightly warmed up. For this reason Mark Buchanan, in his book, compares neo-classical economists to meteorologists that insist in forecasting the weather neglecting storms and hurricanes: the analogy between economics and meteorology is rather stringent, precisely because atmospheric turbulences seem to have much in common with the ups and downs of financial markets and offer interesting ideas to understand the limitations of the hypothesis of economic stability. Moreover, this parallel clarifies the sense in which we may formulate the concept of forecast for a complex systems and how the progresses achieved in the last century allowed improving the predictive power of meteorologists.

One century ago, weather forecasts were based on the analogy with the regularity displayed by several physical phenomena: the idea was simply to try and find in the past a situation sufficiently "close" to the present, and from this draw a forecast for tomorrow. The results of such forecasts were rather disastrous for a reason we now know very well: the atmosphere behaves in a chaotic manner and tiny variations in the physical parameters may induce huge changes in the weather. A breakthrough in weather forecast was achieved thanks to the intuition of the physicist Lewis Fry Richardson, who proposed to employ the equations of the well-known physical laws that rule the dynamics of fluids [5]. Thanks also to the development of computers that allow for the numerical solution of systems of equations, and to the monitoring of the atmospheric conditions through a vast network of satellites, Richardson's ideas have come true and the quality of weather forecast steadily increased in time since the eighties onwards. For instance, making sufficiently reliable seven-day forecasts has become possible starting from the year 2000, while five-day forecasts have nowadays the same reliability as three-day forecasts had in the early nineties.

A recent study [6] has shown that the major economic analysts and the official national and international organizations, besides agreeing all the time, have not been able to foresee, one year in advance, almost none of the 88 recessions (decrease of the real GDP on the yearly basis) that took place in the lapse 2008–2012 in developed countries, so that the authors concluded that "the record of failure in the prevision of recessions is to all extent spotless". This means that nowadays—exactly as in the days before the crisis of ′29—it is a good rule to consider that the contrary of official forecasts, as well as of those of the major analysts, is very likely what will happen. Therefore, it is not surprising that those very same economic models that did not even take into account the possibility of a historical crisis, like that of the year 2008, are not sufficiently reliable to foresee the recessions: in other words, they do not withstand the basic test of every scientific theory.

Neoclassical economics is nowadays equipped with a mathematical guise, as it were a natural science, but is not apt to describe reality, as the failure of all previsions clearly shows: to remedy this aspect they say, obviously a posteriori, that the failures are due to external shocks (e.g., political crises, earthquakes, and so on) that are not included in the models [7]. Nonetheless, comparison with reality is the strength of the scientific method. In physics, for example, several examples can be found of theories that were mathematically correct but completely irrelevant, as they were based on wrong hypotheses. Therefore, such theories yield formally correct results that are contradicted by the experiments. But if an experiment disagrees with a theory, one cannot conclude that this disagreement discredits the quantitative method, rather the hypotheses upon which the model is based should be analysed and those that are wrong must be identified and rejected. And, obviously, the model is changed: more than mathematical accuracy, what matters is physical relevance, that is comparison with reality.

One very important conceptual step, not yet incorporated in neoclassical economics, is the one that gave birth to the interdisciplinary field of the so-called complex systems. Phil Anderson, Nobel prize laureate in physics, synthesized this conceptual revolution in an article of 1970, entitled “More is different” [8]. The basic idea is described hereafter. The traditional approach in physics considers the simplest systems and studies them in detail. This approach, called reductionist, focuses on the elementary building blocks constituting matter and is successfully applied to several phenomena. Thence, it was possible to derive general laws that extend from the scale of the atomic nucleus to the scale of galaxies. However, as soon as the degree of complexity of the structures and systems increases and when these are composed of several interacting elements, one is faced with the problem of following the evolution and the connections of an increasing number of dynamical variables, those describing the behaviour of the individual constituents (for instance, particles, atoms, planets, and so on).

Pursuing the description of the structure as a whole along this route often appears to be helpless. The point is that, when these constituents combine nonlinearly, they form complex structures and support collective motions that are too complicated to be described in terms of the properties of the isolated constituents, due to the huge number of variables and relations involved. We can rather represent this situation as the study of the “architecture” of matter and nature that depends one way or another on the properties of the building blocks, but possesses peculiarities and fundamental laws that cannot be easily connected to those of the single constituents, and are more sensibly described by collective variables. According to Anderson, reality then has a hierarchic structure and at every level of the hierarchy concept and ideas have to be introduced that differ from those employed at the underlying level. In simple words: from the knowledge of the fundamental laws that rule the interactions between elementary particles is not straightforwardly possible to understand the formation of many phases of condensed matter, and new relevant variables, representing the behaviour of several correlated constituents, should be employed. The study of complex systems thus concerns the so-called emergence of collective properties in systems with a large number of constituents in mutual interaction. The definition of the relevant collective variables is not always straightforward and may often require a good deal of ingenuity, and nonetheless this approach has now become fundamental in the study of many physical problems.

Several examples in the last thirty years have shown that understanding some problems specific in a certain field can give rise to a new methodology, possibly applicable to other disciplines, among which, for sure, economics. The study of complex systems thus provides a variety of new and refined methods, that allow for new questions to be formulated and framed in a different and original manner.

An economic theory that does not take into account the way in which the collective behaviour of agents arises or neglects the considerable dependence on small perturbations—as indeed does neoclassical economics—is not apt to explain how important crises or sudden fluctuations, that are seen every day on financial markets, occur. The very idea that an interconnected and tightly interdependent system, like modern financial economy, may tend to some form of stability, must be called into question in a laic and pragmatic manner.

To understand the reach of the theoretical ideas developed about nonlinear phenomena (far) off equilibrium, it is enough to consider that neoclassical economists have interpreted the crisis of 2008 by means of the ideological prejudice according to which the financial crisis was triggered by unforeseeable causes, the bankruptcy of Lehman Brothers, but, since the free markets tend to stability, there would not be consequences on real economy [9]. This interpretation—that influenced the public opinion and the subsequent political choices of the governments of several countries and of the main international institutions—stems from the very theoretical belief according to which deregulated markets should be efficient and rational agents should rapidly adjust every price that is not fully appropriate and every evaluation error. The price should therefore faithfully reflect the underlying reality and ensure the optimal allocation of resources. These "balanced" markets should be stable: therefore crises can only be triggered by large exogenous perturbations, like hurricanes, earthquakes, or political unrests, but certainly not caused by the market itself.

These theoretical prejudices stem from an oversimplification of the problem, so that the idealization not only is unlike reality, but actually is completely inadequate to its understanding. Physicist, who deal with complexity, have been studying since about twenty years systems that display intermittent behaviours much alike those of financial markets, in which the nontrivial nature of the dynamics stems from intrinsic collective effects—not external shocks or causes. The individual parts behave in a relatively simple way, but the interactions lead to collective phenomena, so that the behaviour of the whole system is not simply related to that of its elementary constituents. Even if an equilibrium state exists in theory, this may be totally irrelevant to all practical extent, because the time needed to reach it is too long and because these systems may be intrinsically fragile with respect to the action of tiny perturbations, and evolve in an intermittent way, with a succession of stable epochs spaced out by sudden and unforeseeable changes. Within this perspective, it is natural to conclude that other crises like that started in 2008 may happen again, without any forewarning and, alas!, rather often, as long as a regulation of financial market is not enforced, acting on the endogenous causes of crises and on the theoretical prejudices at the basis of the ineffable equilibrium of free markets.

The relevant question that should be asked is the following: are the fundamental axioms employed in neoclassical economic theory subject to empirical tests? For instance: do the free markets tend to equilibrium, or do they fluctuate wildly? Does the answer to this question come from observations or is it an indisputable assumption? This is a crucial point, since those who think that free markets are efficient and self-regulate toward stable equilibrium will be led to propose an ever more important role of markets and to "starve the beast", the State, corrupted and nepotistic, as indeed happened both in the United States and in Europe in the last twenty years. Those who instead think that free market are dominated by wild fluctuations and are intrinsically far from a stable equilibrium, generating dangerous unbalances and inequalities, will be led to propose a more important participation of the State, trying to improve the efficiency of this latter.

Analogously, those who believe in the stability of deregulated economy will obviously be led to consider every fluctuation in financial markets as an un-influential perturbation of the equilibrium condition: a devastating crisis cannot be foreseen because it is not even conceived in the fairy-tale world of efficient markets.

When a paradigm becomes so strong as to replace whatever empirical observation, it becomes a dogma and the scholar ends up living in the model, without being aware of what happens in real world. As it is well experienced nowadays, there is no stability in real economy: the acceleration toward financialization, with the introduction of derivatives in the market, has made the situation even more potentially explosive. The contrary of what the neoclassical economists do believe.

Queen Elizabeth was then completely right when—during a visit to the London School of Economics in November 2008—asked why the overwhelming majority of economists, those who work in the national and international institutions, and those who write every day on the major newspapers across the world, had not understood that the crisis was going to burst out [10]. Two notorious British economists answered the "Queen's question" with a letter to the Queen, summarizing the positions that had emerged during a forum promoted by the British Academy: "To conclude, in short, Her Majesty, the ineptitude to foresee times, magnitude, and seriousness of the crisis, and to prevent it, although having several causes, was mainly a failure of the collective imagination of many brilliant persons, in this and other countries, to understand the risks for the system as a whole". More explicitly, another group of British economists emphasizes that "in the last years economy became almost entirely a branch of applied mathematics, and got separated from the institutions of the real world and from the events". The problem does not stand in the question whether economics is or is not an exact science (and sure it is not), or in the fact that the use of mathematics provides a solid scientific appearance, rather it is methodological.

Biology (which is not an exact science too) has made significant progresses in the last years, thanks to a systematic analysis of experiments and data, advancing in a pragmatic manner, and not being guided by indubitable ideological assumptions.

The economic crisis initially started as a banking and financial crisis, triggered by a crisis of the private debt, due to an uncontrolled creation of "money out of nothing" in the form of derivative bonds on the banks' side, both in Europe and America, justified by the belief that they would have more effectively stabilized the markets. When this house of card crumpled, with enormous costs for millions of individuals, the American government upheld the banks with almost thirty trillion dollars, in the form of credits and guarantees, some returned back, some not, while at the end of 2010 the European Commission authorized support to the banks for more than four trillion dollars. With these interventions the financial crisis, that until the beginning of 2010 had been a crisis of private banks and had not turned into a worldwide catastrophe, was charged onto the public balances. At that moment, the leitmotiv, diffused by the major media, often by the same neoclassical economists that had supported in the past every choice of deregulation and liberalization of the markets, turned different: excess of indebtedness of the United States, excess of public expenditure, unbearable pensions, expenses for education that "we can no longer afford", and so on. The message is put forward that the State spends too much and therefore it is necessary to cut the public expenses: kindergartens, schools, healthcare, education, research, pensions, etc.

This incredible and unacceptable mystification has been made possible thanks to a surprising and pervading cultural hegemony, that developed not only by means of the conquest of dominant academic positions achieved by neoclassical economists, but also, mainly, by means of the superposition of their academic, political, and opinion-making roles, as many of them were columnists of major national newspapers or the prince's advisors. Some theoretical ideas that appear to be innocent cogitations of some eccentric scholar, have become powerful means of political and cultural brainwashing. The Nobel laureate in economics Paul Samuelson writes: "I do not care about those who write the laws of a nation or elaborate her treaties, as long as I can write her textbooks of economics". The way out of the economic crisis passes first of all through a change of cultural perspective that the new concepts of natural sciences, together with their methodologies, can contribute to economics.

To conclude, we wish to discuss the intellectual honesty of neoclassical economists as related to the way in which they argue and uphold the unbelievable hypotheses at the basis of models currently employed by political and institutional decision-makers. To this purpose, hereafter we report the illuminating lecture of the Nobel laureate in physics Richard Feynman about the meaning of science and scientific ethics [11]. In our opinion, all the premises and elements that are necessary to identify the methodological flaws of neoclassical economics can be found in Feynman's *incipit*:

> *In the South Seas there is a cargo cult of people. During the war they saw airplanes with lots of good materials, and they want the same thing to happen now. So they've arranged to make things like runways, to put fires along the sides of the runways, to make a wooden hut for a man to sit in, with two wooden pieces on his head to headphones and bars of bamboo sticking out like antennas—he's the controller—and they wait for the airplanes to land. They're doing everything right. The form is perfect. It looks exactly the way it looked before. But it doesn't work. No airplanes land. So I call these things cargo cult science, because they follow all the apparent precepts and forms of scientific investigation, but they're missing something essential, because the planes don't land.*

This metaphor of the cargo cult of people, put forward by Feynman to criticize all those scientists that stick to the appearances and the formal aspects, without digging deep into the essence of things and facts, in our opinion, applies equally well to the neoclassical economists. Let us follow Feynman's line of reasoning further on:

> *Now it behaves me, of course, to tell you what they're missing. But it would be just about as difficult to explain to the South Sea islanders how they have to arrange things so that they get some wealth in their system. It is not something simple like telling them how to improve the shapes of the earphones. But there is one feature I notice that is generally missing in cargo cult science. That is the idea that we all hope you have learned in studying science in school—we never say explicitly what this is, but just hope that you catch on by all the examples of scientific investigation. It is interesting, therefore, to bring it out now and speak of it explicitly. It's a kind of scientific integrity, a principle of scientific thought that corresponds to a kind of utter honesty—a kind of leaning over backwards. For example, if you're doing an experiment, you should report everything that you think might make it invalid—not only what you think is right about it: other causes that could possibly explain your results; and things you thought of that you've eliminated by some other experiment, and how they worked—to make sure the other fellow can tell they have been eliminated.*

Ethics, under the form of scientific integrity and intellectual honesty, is Feynman's main concern. Severity in the scrutiny of the outcomes of experiments on how the objects we aim to describe do really behave in the world is a necessary condition for a discipline to be termed scientific. Let us now peruse the conclusion of Feynman's lecture:

> *[…] We've learned from experience that the truth will come out. Other experimenters will repeat your experiment and find out whether you were wrong or right. Nature's phenomena will agree or they'll disagree with your theory. And, although you may gain some temporary fame and excitement, you will not gain a good reputation as a scientist if you haven't tried to be very careful in this kind of work. And it's this type of integrity, this kind of care not to fool yourself, that is missing to a large extent in much of the research in cargo cult science. […] So I have just one wish for you—the good luck to be somewhere where you are free to maintain the kind of integrity I have described, and where you do not feel forced by a need to maintain your position in the organization, or financial support, or so on, to lose your integrity. May you have that freedom.*

Feynman's "cargo science" is therefore a perfect metaphor of the neoclassical theory: the runways, the fires and the bars of bamboo are much alike the invisible hand, the rational expectation and the efficient markets. The same way as the "cargo people" are waiting for the airplanes to land, the economists wait for the markets to stabilize. Unfortunately for the South Seas islanders and for us consumers and citizens the airplanes are not landing and the crises keep on raging. Neoclassical theory probably lacks something essential: complexity is certainly one of the most promising attempts to provide scientific instruments, and reformulate from the very foundation the methodological approach but one should never forget that economics is too strictly related to politics to be studied as a natural science.

Bibliography

1. Buchanan, M. (2014) Forecast: What Physics, Meteorology, and the Natural Sciences Can Teach Us About Economics, Bloomsbury USA; Reprint edition (June 3, 2014) [Italian Edition: Mark Buchanan, Previsioni. Cosa possono insegnarci la fisica, la meteorologia e le scienze naturali sull'economia, Malcor D'E (2014)]
2. Sylos Labini, F.: Science and the Economic Crisis: Impact and Lessons. Springer (2016)
3. Gillies, D. (1993) Philosophy of Science in the Twentieth Century: Four Central Themes. Blackwell (1993)
4. Bouchaud, J.-P.: The (unfortunate) complexity of the economy. Physics World, pp. 28–32 (2009)
5. Hunt, J.C.R.: Lewis Fry Richardson and his contributions to mathematics, meteorology, and models of conflict. Ann. Rev. Fluid Mech. **30**: xiii—xxxvi (1998). See also A. Vulpiani, Lewis Fry Richardson: scientist, visionary and pacifist. Lettera Matematica (International Edition), September 2014, pp. 121–128 (2014)
6. Ahir, H., Loungani, P.: Fail Again? Fail Better? Forecasts by Economists During the Great Recession, George Washington University Research Program in Forecasting Seminar (2014)
7. Gillies, D.: The use of mathematics in physics and economics: a comparison probabilities, laws, and structures. Philos. Sci. European Perspect. **3**, 351–362 (2012)
8. Anderson, P.W.: More is different. Science **4**, 393–396 (1972)
9. Bouchaud, J.-P.: Economics needs a scientific revolution. Nature **455**, 1181 (30 October 2008)
10. Gillies, Donald: Economics and research assessment systems. Econ. Thought Pap. Rev. **1**, 23–47 (2009)
11. Feynman, R.: Cargo cult science (Engineering and science) **37,** 7 (1974)

Bibliography

Research Habits in Financial Modelling: The Case of Non-normality of Market Returns in the 1970s and the 1980s

Boudewijn de Bruin and Christian Walter

Abstract In this chapter, one considers finance at its very foundations, namely, at the place where assumptions are being made about the ways to measure the two key ingredients of finance: risk and return. It is well known that returns for a large class of assets display a number of stylized facts that cannot be squared with the traditional views of 1960s financial economics (normality and continuity assumptions, i.e. Brownian representation of market dynamics). Despite the empirical counterevidence, normality and continuity assumptions were part and parcel of financial theory and practice, embedded in all financial practices and beliefs. Our aim is to build on this puzzle for extracting some clues revealing the use of one research strategy in academic community, model tinkering defined as a particular research habit. We choose to focus on one specific moment of the scientific controversies in academic finance: the 'leptokurtic crisis' opened by Mandelbrot in 1962. The profoundness of the crisis came from the angle of the Mandelbrot's attack: not only he emphasized an empirical inadequacy of the Brownian representation, but also he argued for an inadequate grounding of this representation. We give some insights in this crisis and display the model tinkering strategies of the financial academic community in the 1970s and the 1980s.

To the extent that the Global Financial Crisis is seen as a moral crisis, most commentators consider it to be a crisis caused by an elite of highly educated finance professionals relentlessly and excessively pursuing their own interests. They see the crisis as a crisis of greed. And when politicians, policymakers and private citizens call for the restoration of trust in finance, the primary target is the bonus culture, in which bankers are lavishly compensated if upside risks manifest, but where the tax payer

B. de Bruin
Faculties of Philosophy, and Economics and Business, University of Groningen,
Nettelbosje 2, 9747 Groningen, AE, The Netherlands
e-mail: b.p.de.bruin@rug.nl

C. Walter (✉)
Fondation Maison des Sciences de L'homme, Ethics and Finance Chair,
190 Avenue de France, 75013 Paris, France
e-mail: christian.walter@msh-paris.fr

E. Ippoliti and P. Chen (eds.), *Methods and Finance*,
Studies in Applied Philosophy, Epistemology and Rational Ethics 34,
DOI 10.1007/978-3-319-49872-0_5

covers the downside risks. If trust is what we want to restore, then it is important to see that intelligently placed trust [1] depends on trustworthiness. Trustworthiness, not trust, has to come first.

Following a simple theory, trustworthiness depends on two components: the trusted person's motivation, and his or her competence. You trust your doctor to the extent that she is willing and capable to do the work she has to do. Using this theory, one might say that so far commentators and policymakers have primarily focussed on the motivation component of the financial services industry: culture programmes—which are being implemented in almost any bank today—are programmes aimed at getting employee motivation right. But if that is true, then we run the risk of doing too little to boost competence in the industry.

This brings us to a second observation about the way most commentators see finance. Bankers may be depicted as greedy, but they are hardly ever seen as incompetent. Quite to the contrary, many of them are portrayed as highly skilled quantitative geniuses that work on products that one can only understand if one has a PhD in physics. The *quants*, as they are sometimes called, may have all sorts of immoral motivations to do the work they do, but at least they are clever. Or so most people believe.

That belief is wrong. And this has serious consequences. Economists [2, 3], sociologists [4] and philosophers [5] have convincingly demonstrated widespread lack of competence, a failure to practise open-mindedness and a genuine aversion against evidence-based methodologies in many parts of the financial sector, the regulatory and supervisory authorities included [6, 7]. The credit rating of structured securities is a case in point [8], and it is quite likely that the subprime mortgage crisis would have hit with much less force if mortgage-backed securities had been rated in more rigorous and methodologically sophisticated ways [9, 10]. One year after the crisis, a leading global player in the asset management industry acknowledged that 'the primary reason for this underestimation of risk *lies in the conventional approach to applying mean-variance theory*, which was pioneered by [the Nobel Prize winner] Harry Markowitz in 1952' [11, 3, our emphasis]. The impact of this underestimation of risk was huge: 'we believe that conventionally derived portfolios carry a *higher level of downside risk than many investors believe*, or current portfolio modelling techniques can identify' [11, id., our emphasis].

This chapter is motivated by a concern for competence in finance. But unlike the above-reference publications, one build on the acknowledgement of past mistakes given in [11] and one considers finance at its very foundations, namely, at the place where assumptions are being made about the ways to measure the two key ingredients of finance: risk and return in the mean-variance framework. Despite such exceptions as Bachelier's [12] early work on asset returns, financial practice was up to the 1960s informed by what can be called *folk theories* telling practitioners how to value securities as described in Bernstein's memories [13]. Technical analysis or *chartism* worked hand in hand with the analysis of fundamentals and a motley of not-too-complicated mathematical techniques [14–16]. Inspired by probability theory [17] and with the emergence of increased availability of empirical data [18], Harry Markowitz [19], William Sharpe [20] and Eugene Fama [21, 22] started developing a rigorous and novel approach to financial economics qualified by Mehrling [23] as a 'revolutionary idea of finance'. Large databases were created, stimulated by the increased relevance of computer technology, with the Chicago Center for Research in Security Prices (1960) as the most important academic player. And gradually, this new approach to finance gained hold of business school curricula as well.

Ironically, however, it did not take too long for discrepancies between theory and empirical reality to appear. Returns for a large class of assets display a number of *stylized facts* [24] that cannot be squared with the traditional views of 1960s financial economics. This traditional view, arisen in the 1930s, is to the effect that return follow Gaussian random walks, where a random walk is a stochastic process in discrete time in which increments are independent and identically distributed (IID). If the distribution is Gaussian, it is a Gaussian random walk; if the distribution is, say, a Gamma distribution, it is a Gamma random walk, and so on.

Already in the 1950s, empirical studies revealed some of the problems of this view. It appears that the empirical distributions of returns are *leptokurtic* (from the Greek words *leptos*, peaked, and *kurtosis*, curvature); that is, they are more peaked that the Gaussian bell, exhibiting fat tails and values clustered around the mean, with the result that extreme events are more likely than under a normal distribution. They are also asymmetric or negatively skewed in the sense that it is particularly negative extreme events that are more likely. And the paths of returns also display volatility clustering, which means that large changes tend to be followed by large, and small by small changes. These stylized facts, which are clearly a violation of the IID hypothesis, have long been known in the academic community. In a 1953 landmark paper Kendall noted that the results from price data between 1883 and 1934 appear 'rather leptokurtic' [25, 13], and this view was reinforced in later publications during the 1950s and early 1960s by many authors (see below).

Despite the empirical counter evidence, normality and continuity assumptions are part and parcel of financial theory and practice, even today. They form the common core of mean-variance analysis, Black-Scholes-Merton option price theory, and fundamental asset pricing methods that emerged from the work of Harrison, Kreps, and Stanley between 1979 and 1981. They are widely used in Value-at-Risk (VaR) analysis in banks and insurance companies. For instance, the so-called *square-root-of-time-rule* for calculating minimum capital underlying such regulatory requirements as Basel III and Solvency II is a very narrow subset of time scaling rule of risk, and comes directly from a Gaussian framework committed to IID. Even though financial modelling progresses in non-Gaussian and non-Brownian directions,[1] the bulk of business school finance is still traditional, supporting the myth that risk can be completely tamed. But, 'using standard deviation, rather more behaviorally attuned conditional VaR measures, may in fact inadvertently *increase rather than decrease* downside risk' [11, 6, our emphasis].

In this chapter we tentatively exploit an interesting analogy between financial modelling and another respected branch of mathematical economics: game theory. Game theory studies strategic interaction between economic agents, applying and

[1]See, for example among others, [26–35].

extending results from probability theory and topology. But just as finance, it has suffered a rather serious clash with empirical reality: the behaviour of real players is often very different from what the Nash equilibrium concept or its refinements predict. In a series of publications, De Bruin [36–39] has interpreted this episode in the history of economic thought as resulting from the adherence to a number of fruitless research habits that stem from an uncritical adoption of the view that the social sciences need only be true in the abstract. These research habits include overmathematization, the use of intuitions as data, introversion, model tinkering and instrumentalism. To be sure, some of these habits were much less prominent in financial modelling. Unlike game theory, financial economists have from the very start been concerned with data—not seldom was their main research aims the rather practical preoccupation to help the financial industry. Moreover, the history of financial modelling seems more complex and multifaceted than that of game theory. In a series of publications related to the history of financial thought, Walter [40–43] has produced an history of financial modelling and presented the intertwinings between financial modelling and financial practices, exhibiting the role of the random walk model as a backbone of the financial modelling in the twentieth century. The notion of 'leptokurtic phenomenon' [44] has been introduced at the heart of this history to isolate a crucial moment for financial modelling issues: awareness of 'anomaly', i.e. the recognition that the effective behaviour of markets has violated the Bachelier-Osborne paradigm of Brownian representation. This moment was that of the birth of hard scientific controversies concerning the nature of the dynamics of markets. As Kuhn said, 'confronted with anomaly or with crisis, scientists take a different attitude toward existing paradigms' [45, 91]. As a result our rather more modest aim in the present chapter is to build on these works for extracting from the 'leptokurtic crisis' of the 1970s some clues revealing the use of one research strategy in particular—model tinkering. This research habit allows to 'normal science' (in Kuhn's notion) to handle the crisis-provoking problem despite the despair of those who have seen it as the end of an existing paradigm' [45, 84].

1 Model Tinkering and Research Habits: A Case Study

The methodology we employ here was developed in a series of papers in which De Bruin compared two research programmes in game theory, a branch of mathematical economics the stated aim of which it is to explain the strategic behaviour of interacting rational agents [36–39]. Summarizing some of these materials, we here introduce the methodology and some of the results to the extent that our argument about financial modelling depends on it.

While a number of nineteenth century economists count among the precursors of game theory, the field started with the publication of *Theory of Games and Economic Behavior* in 1944, written by John von Neumann, a mathematician, and Oscar Morgenstern, an economist that would also write a number of important papers in mathematical finance [46]. Their book developed novel mathematical tools to examine strategic interaction between economic agents, stimulated several generations of researchers to work in a field that would soon be known as *game theory*. An important role in this field would be played by John Nash. A student of von Neumann's—the film *A Beautiful Mind* recounts his life—Nash developed an equilibrium concept describing rational interaction in one-shot situations. Here is how he begins:

> We proceed by investigating the question: what would be a 'rational' prediction of the behavior to be expected of rational [*sic*] playing the game in question? By using the principles that a rational prediction should be unique, that the players should be able to deduce and make use of it, and that such knowledge on the part of each player of what to expect the others to do should not lead him to act out of conformity with the prediction, one is led to the concept of a solution defined before [47, 23].

From these assumptions, Nash developed his eponymous equilibrium solution concept. Yet the Nash equilibrium was soon found to be 'intuitively unreasonable' [48] or to deliver results 'inconsistent with our intuitive notions about what should be the outcome of game' [49]—criticisms that started the quest for 'refinements' of the Nash equilibrium, that is, for alternative solution concepts that would keep 'right' and eliminate 'irrational' outcomes. Some examples include the subgame-perfect equilibrium [50], the perfect equilibrium [48] and the proper equilibrium [49]. The Nash Equilibrium Refinement Programme (NERP) gained great influence within the field. All respected textbooks in the field contain elaborate treatments of the various refinements, Nobel Prizes were awarded to Nash and two refiners (John Harsanyi and Reinhard Selten), and as one commentator said, 'the equilibrium concept of Nash...together with its refinements, is without doubt the single game theoretic tool that is most often applied in economics' [51, 1].

It was, however, clear from the very start of NERP that the Nash equilibrium suffered not just from the fact that it was incompatible with certain intuitive notions of rational strategic interaction; rather plain empirical counter evidence was available too. NERP solutions were criticized for failure adequately to describe experimental and empirical results [52], as well as for their being unable to ground their attempted explanations in the processes that were thought to be ultimately explanatory: the players' rationality, that is, their maximizing their utility given their probabilistic beliefs. These two factors—let us call them *empirical inadequacy* and *inadequate grounding* —were among the reasons that led to the development of a different strand of game theoretic research: the Epistemic Programme. Firmly committed to explaining strategic interaction on the basis of players' beliefs, utility functions and rationality, researchers in this programme used methods from measure theory and topology, pursuing a research agenda centred around the concept of *interactive rationality*. The Epistemic Programme is not the only available alternative to NERP; behavioural [52], evolutionary [53] and stochastic [54] methods have been proposed as well. As witnessed by the number of recent publications devoted to it in such journals as *Econometrica* and *Games and Economic Behavior* as well as by the importance it has gained in recent textbooks, however, the Epistemic Programme seems to have taken centre stage [55, 56].

What might explain the relative success of the Epistemic Programme? De Bruin [36–39] has shown that a conceptually and historically fruitful way to distinguish

NERP and the Epistemic Programme focusses on a number of research habits that characterize NERP, and we draw from these publications to give a brief introduction here.[2] An early source of inspiration here is John Stuart Mill's 1836 article 'On the definition of political economy; and on the method of philosophical investigation in that science' [57]. Mill there introduces the distinction between *a posteriori* and *a priori* methods in science, that is, the distinction between inductive methods that start with experience, on the one hand, and deductive methods that start with abstract views about the world, on the other; and he observes that what sets these methods apart is the possibility of using an *experimentum crucis*, a decisive experiment to bring the theory to the test. Mill maintains that while this is possible in the natural sciences, the social sciences do not so easily allow for decisive experiments because of the sheer number of causal factors and because of the difficulties attached to replication of experiments. Unlike the inductive natural scientist, the deductive social scientist, according to Mill, therefore resides to a process in which observations are simplified and abstracted so as to deliver statement that 'are true, as the common phrase is, *in the abstract*'. Some causal factors will, in the process of abstraction and simplification, find no representation in the model. When applied to a concrete case at hand, these 'disturbing causes' that 'have not fallen under the cognizance of science' can, and must, be included. This makes sense, Mill thinks, because 'that which is true in the abstract, is always true in the concrete with proper *allowances*'. What in the end makes social sciences different from natural sciences is the presence of disturbing causes.

This *true-in-the-abstract* view of social science finds an echo in the attitudes several game theorists—and economists, more broadly—adopt towards their field. Robert Aumann, for instance, holds onto the view that game theory is not descriptive 'in the sense that physics or astronomy are' [51]. Rather he believes that game theory describes *homo rationalis*, which, according to him, is a 'mythical species like the unicorn and the mermaid'. In fact, he finds it 'somewhat surprising that our disciplines have any relation at all to real behavior', and that we can gain '*some* insight into the behavior of *Homo sapiens* by studying *Homo rationalis*' [51, 36]. This is a version of Mill's true-in-the-abstract view, albeit a rather extreme one. The relevance in the context of the present discussion is that a social scientist holding it is, we believe, susceptible to a number of research habits. To begin with, the true-in-the-abstract supports *mathematization* of social science narratives, with publications in economics and other social science journals hardly differing from papers in mathematics with their conformance to Bourbakian templates of definitions, theorems, proofs. Mathematization as such is not, of course, put into question. What is at stake is, however, the relative amount of attention paid to mathematics at the expense of scientific input and inspiration from such fields as psychology, cognitive science, sociology or anthropology, which is reflected in the fact that only a small proportion of references in economics are to publications outside economics, as opposed to such fields as sociology.

[2]See, in particular, [39, 128–134].

A second research habit that a true-in-the-abstract view leads to is the appeal to the researcher's private *intuitions* to support particular modelling assumptions. Without too much exaggeration one might say that this habit lies at the very bottom of the entire neo-classical approach to economics, dependent as it is on the concept of preference and utility as von Neumann and Morgenstern developed it. Myerson asks the question of '[w]hy should I expect that any simple quantitative model can give a reasonable description of people's behavior?' And he answers it by noting that '[t]he fundamental results of decision theory directly address this question, by showing that any decision-maker who satisfies certain *intuitive axioms* should always behave so as to maximize the mathematical expected value of some utility function, with respect to some subjective probability distribution' [58, 5, emphasis added]. Intuitions keep the research in the realm of the true-in-the-abstract; they are easier to generate than empirical data; they allow researcher to sidestep econometric subtleties that typically plague experimental researchers; they are more difficult to refute than experimental data; they are obtained at no cost—and they are expressed verbally, with all the inherent vagueness and room for rhetorics and intended ambiguity. Just as mathematization, an appeal to intuitions is not as such at odds with the aims of social science. What should be a reason to worry is when intuitions replace experiments where experiments could have carried out.

An appeal to intuitions instead of empirical data is often accompanied by *introversion*, that is, the research habit of excessively focussing on internal and often technical problems instead of on empirically or experimentally motivated issues. This habit comes to the fore very clearly again in the way in which Aumann, the game theorist, explains the attractions of the principle of utility maximization, of which, according to him, the validity 'does not depend on its being an accurate description of true individual behavior'; instead, it 'derives from its being the underlying postulate that pulls together most of economic theory', or in other words, '[i]n judging utility maximization, we must ask not "Is it plausible?" but "What does it tie together, where does it lead?" [51]. This, by the way, also functions as a theoretical tool in Aumman's attack on Herbert Simon's [59] more empirically adequate principle of satisfying. Such and similar alternatives to the principle of utility maximization 'have proved next to useless in this respect', Aumann maintains, because '[w]hile attractive as hypotheses, there is little theory built on them; they pull together almost nothing; they have few interesting consequences' [51].

Introversion also shows when researchers tend to overinterpret the models they make. When this happens, what counts for a researcher is not just whether the model adequately describes a particular phenomenon, but rather whether all and every aspect of the model has a potential interpretation; and if they do not, then this is considered paradoxical. Certain games, for instance, that do not seem to model any real-life strategic interaction receive excessive attention. Relatedly, introversion can be witnessed in the way in which researchers set aside or describe certain results as 'undesired'. Rather than seeing empirical falsification as the ultimate counterargument against a theoretical construct, their main worry seems to be whether the construct is compatible with existing theory. Introversion, then, sanctions a rather conservative attitude in science.

The last research habit or attitude associated with true-in-the-abstract conceptions of economics is *instrumentalism*, which was most famously defended by Milton Friedman [60]. According to Friedman, what matters about a theory is whether it predicts well, not whether it makes realistic assumptions. Instrumentalism, however, risks lowering explanatory ambitions. An oft-quoted example is this:

> Consider the density of leaves around a tree. I suggest the hypothesis that the leaves are positioned as if each leaf deliberately sought to maximize the amount of sunlight it receives, given the positions of its neighbors, *as if* it knew the physical laws determining the amount of sunlight that would be received in various positions and could move rapidly or instantaneously from any one position to any other desired and unoccupied position [60].

The density of leaves may be adequately described by Friedman's theory, but if that is all we care about, the need to do empirical research on plants becomes less pressing, and the likelihood of discovering related phenomena smaller. How would an instrumentalist, for instance, discover that the plants respond to *blue* but not red light? The instrumentalist adopting a true-in-the-abstract 'as-if' view cannot, moreover, explain the difference between human, animal and plant agency. He or she cannot avoid vacuity or multi-realizability of explanations. For instance, a typical human action is compatible with a whole range of combinations of probability distributions and utility functions. An instrumentalist does not care which of them is the 'right' one. Where theories are developed for purely instrumental reasons and need to be true in the abstract only the adequacy of the representation of the underlying causal structures and processes matters much less; and where theory and reality clash, the method of model tinkering has, if we adopt such a view, much to recommend, for instead of revising the theory where it hurts, mathematical epicycles can be added to the model resolving the conflict with the data'—and it does not matter whether these are empirical, or based on intuition only.

It is true that *ad hoc* models can be found in physics and other scientific disciplines as well; the best-known such model is indeed still Ptolemy's theory of planetary motion. But physicists would generally try to avoid introducing theoretic terms and statistical parameters that cannot be given a physical meaning. The theories of secondary school physics do not correctly describe the trajectories of balloons and other exceedingly light objects. Including air resistance as a causal determinant makes the description more accurate. The increased descriptive accuracy is not, however, the only reason to embrace the more complex theory; another is that it uncovers a reality that was so far hidden: the impact air has on the movement and acceleration of falling objects.

2 Financial Modelling

Intuitions as data, introversion, instrumentalism—we argue now that these research habits played a significant role in shaping developments in financial modelling in the second half of the twentieth century. We first explain the standard model of market returns, the exponential Brownian motion, which in a sense was the Nash equilibrium of financial modelling. Subsequently, we zoom in on a number of mutually competing programmes (mixed diffusion jumps processes, ARCH modelling, stochastic volatility, infinite activity, tempered stable processes and Mandelbrot's programme). Given the space we have, we cannot hope to cover all details of this complex and extensive territory, and hence we aim at pointing out a number of events in the history of financial modelling making our main claim reasonably plausible. We do presuppose a fair amount of knowledge of financial modelling here, as we see philosophers of science and financial mathematicians as our primary audience.

2.1 Standard Model

Let us denote the price of any financial asset (stock, bond or other) at time t by $S(t)$. Practitioners (traders, risk managers, etc.) are generally interested in the cumulative continuous rate of return between times 0 and t, $X(t) = \ln S(t) - \ln S(0)$. The classic Brownian motion representation of return dynamics, due to Louis Bachelier [12] and Maury Osborne [61] is given by

$$X(t) = \mu t + \sigma W(t), \tag{1}$$

where $W(t)$ is a standard Wiener process, μ a trend parameter and σ a volatility parameter meant to capture what is thought as market risk in the classical paradigm of Markowitz-Sharpe (portfolio theory) and Black-Scholes-Merton (option pricing theory). This model became the standard model of market dynamics in 1965 with Paul Samuelson. This Eq. (1) has three important consequences. First, it supports a Gaussian distribution with mean μt and variance $\sigma^2 t$ for the marginal law of empirical returns $X(t)$. Secondly, it underscores the idea that $(X(t), t \geq 0)$ is a stochastic process in which the increments are independent and identically distributed, the IID hypothesis we encountered earlier. And thirdly, it entails time scaling of distributions—and consequently time scaling of risk—in the sense that a given horizon (e.g., t) of a return distribution is scaled to another (e.g., $t \times a$) as in

$$X(t \times a) \stackrel{d}{=} X(t) \times \sqrt{a} \tag{2}$$

where the symbol $\stackrel{d}{=}$ is used to signify equality in distributions. This is called the *scaling property* of Brownian motion or the *square-root-of-time rule* of scaling.

If this is a quick and dirty overview of Brownian motion, it is important to emphasize the widespread use of these models, particularly in the financial industry and for regulatory purposes—and to emphasize its wide empirical inadequacy at the same time. Let us exemplify with the scaling property and its practical implementations in prudential rules. The scaling property (2) leads to scaling of volatility in the sense that

$$\sigma(a \times t) = \sigma(t) \times \sqrt{a} \tag{3}$$

which supports the widely used practice to compute the annual volatility from the weekly one. In this example, $t = 1$ week so $a = 52$ weeks, therefore annual volatility = weekly volatility $\times \sqrt{52}$.

The so-called square-root-of-time rule is widely used in Basel III and Solvency II regulations which support the calculation and the implementation of a probabilistic measure of market risk called 'Value-at-Risk' [62] (hereafter VaR). The minimum capital requirement is an estimated quantile of a return distribution (10 days 95 % VaR metrics). The 10 days VaR is obtained with a simple application of time scaling of risk using the square-root-of-time rule: we have VaR 10 days = VaR 1 day $\times \sqrt{10}$.

2.2 Anomalies with the Standard Model

But the square-root-of-time rule leads to a systematic underestimation of risk that worsens with the time horizon. This is because the scaling property (3) is not corroborated by empirical facts and this fact illustrates the *empirical inadequacy* of the standard model. Among a lot of counter-examples [24], let us give the intuition of scaling refutations. From this simple relationship (3) is derived the statistics of the so-called 'variance-ratio tests': the idea behind is just to compute the ratio 'annual volatility / weekly volatility' and to check if the result equals $\sqrt{52}$ (with appropriate confidence intervals). If is the case, there is corroboration of the scaling law hypothesis. All results are negative and Lo and MacKinlay conclude that 'The random walk model is strongly rejected for the entire sample period (1962–1985)' [63, 41]. Consequently, as Danielson and Zigrand [64] argue, 'even if the square-root-of-time rule has widespread applications in the Basel Accords, it fails to address the objective of the Accords'. VaR scaling does not contribute to the realization of the Basel objectives [6]. The refutation of the $\sqrt{10}$ law explains incidentally the multiplicative factor of 3 in Basel Accords which is intended to compensate for errors that can arise in simplifying assumptions [65, 255]. With this example, we enter the *anomalies*, in terms of *empirical inadequacy*, of the standard model.

Nor is time scaling the only trouble. As we mentioned before, Brownian motion clashes with a number of empirical stylized facts about return distributions. Returns distributions violate normality in that empirical distributions are leptokurtic compared to the normal distribution, and return processes violate the IID hypothesis in the sense that there is serial correlation in squared returns, that is, volatility dependence. Perhaps most importantly, continuity is violated. Merton, for instance, wrote that 'there is a *prima facie* case for the existence of jumps' [66], that is, for discontinuities, and Cox and Ross agreed that 'exploring alternative forms [of motion] is useful to construct them as jump processes' [67]. A decade later when Black-Scholes option pricing models had become widely popular, Ball and Torous pointed out that 'empirical evidence confirms the systematic mispricing of the Black-Scholes call option pricing model', noting that '[t]he Merton model which explicitly admits jumps in the underlying security return process, may potentially eliminate these biases' [68]. And more recently still, Carr and his co-authors wrote that they 'seek to replace this process with one that enjoys all of the fundamental properties of Brownian motion, *except for pathwise continuity and scaling*, but that permits a richer array of variation in higher moment structure, especially at shorter horizons' [69]. Unlike the perhaps more introverted game theorists, the finance academics were well aware of the discrepancies between model and reality. The challenge was how to respond to them.

2.3 Early Refutations of the Brownian Representation

These stylized facts had been known for a long time. As early as the 1950s, empirical studies pointed out the problems of the Brownian representation of market dynamics. Below are some examples of this evidence. We quote rather extensively to point to what is historically and epistemologically an intriguing phenomenon: while statisticians found increasing evidence backing the diagnosis of non-normality of market dynamics, around 1965 finance academics made a conscious choice to ignore the evidence, opting for the standard model of reducing return processes to Brownian motion. In a 1953 landmark paper published in the respected *Journal of the Royal Statistical Society*, Maurice Kendall observed of price data between 1883 and 1934 that '[t]he distributions are accordingly *rather leptokurtic*' [25, 13, emphasis added]. Seven years later, Arnold Larson noted that '[e]xamination of the pattern of occurrence of all price changes in excess of three standard deviations from zero...indicated...presence in the data of an *excessive number of extreme values*' [70, 224]. In a very important article published in 1961 in the *American Economic Review*, Houthakker wrote that

> [t]he distribution of day-to-day changes in the logarithms of prices does not conform to the normal curve. It is...*highly leptokurtic*...The variance of price changes does not seem to be constant over time...Very large deviations, in fact, seem to come in bunches. The non-normality mentioned above may be related to the changing variance [71, 168, emphasis added].

And in the same year, Sydney Alexander emphasized pithily that '[a] rigorous test...would lead to dismiss the hypothesis of normality...This sort of situation (*leptokurtic*) is frequently encountered in economic statistics' [72, 16].

All this concerns only one stylized fact: non-normality. But as early as 1959, Harry Roberts had argued in the *Journal of Finance* against the assumption of IID, stating that

> modern statistical theory has been largely built up on the assumption of independence. Much of it also assumes...that the underlying distribution is a normal distribution in the technical sense of that term. The assumption of normality usually seems far less crucial to the applicability of statistical methods than does that of independence [73, 14].

The assumption of IID was attacked from other sides as well. In an influential publication, Paul Cootner observed that real markets do not follow the random walk model. And while '[t]he way in which actual markets operate is one of the more fascinating of current economic questions', he admitted that '[i]f their behavior is more complicated than the random walk models suggest, it will take more sophisticated statistical testing to discover it' [74, 44].

3 The Leptokurtic Problem

All of which is to say that the 'leptokurtic phenomenon' [44] is not new. It is its institutional acknowledgement which is new. If the non-normality of market returns seems nowadays widely accepted by practitioners and finance academics, it was not the case during almost forty years. We elaborate now this issue with the aim to introduce the leptokurtic crisis opened by Mandelbrot in 1962.

3.1 The Leptokurtic Crisis

The leptokurtic crisis was opened by Mandelbrot in 1962. In a series of pioneering contributions between 1962 and 1973, Mandelbrot tackled the symptoms of non-normality by paving the way for new methods of statistical economics: first by incorporating strongly non normal market returns [75, 76], then by incorporating non Brownian scaling rules with long-term dependence [77] and finally by combining these propositions in three epistemological papers [78–80].

These papers exploded the field of financial modelling and launched violent controversies dividing the community of finance academics into two opposite camps: pro and cons the 'wild randomness' (see below) introduced to solve the leptokurtic puzzle. This crisis started the quest for 'refinements' of the Bachelier-Osborne model, that is for alternative models that could save the 'mild' randomness for solving the leptokurtic problem. Following the de Bruin's approach of game theory, we suggest here that these attempts could be considered as the Bachelier-Osborne refinement programme (BORP), whereas the three papers of Mandelbrot [78–80] could define what we call the Mandelbrot programme. During more than thirty years, these two programmes were incompatible. Mandelbrot and his opponents speak from 'incommensurable' viewpoints in the Kuhn's words. There was no 'neutral statistical test' for model-choice. The profoundness of the crisis came from the angle of the Mandelbrot's attack: not only he emphasized an *empirical inadequacy* of the Brownian representation, but also he argued for an *inadequate grounding* of this representation.

3.2 *Turning Point: Mild Randomness Vs Model Tinkering*

We now use a perspicuous representation of the two tracks within financial modelling as they appear in the 1960s, in order to make the research habits visible.

3.2.1 The Leptokurtic Evidence

We do this by moving to a discrete multiperiod model of securities markets with a finite number of trading dates occurring at time $k \in \mathbf{N}$. This move can be viewed as a way for simplifying the representation. In this framework, Eq. (1) becomes

$$X_k = X_{k-1} + \mu + \sigma u_k, \quad (4)$$

with $u_k \rightarrow \mathcal{N}(0, 1)$ capturing Gaussian white noise with unit variance, and σ the volatility of returns, a constant. Let us write $\varepsilon_k = \sigma u_k$. This quantity 'shapes' the randomness, i.e. the risk of the market dynamics

$$\varepsilon_k = \sigma \times u_k, \quad (5)$$

which expresses the idea that risk equals 'scale' of fluctuations times 'shape' of randomness. The volatility is the scale parameter, the random term defines the morphology of risk. To put it simply, the leptokurtic problem is: ε_k is non-Gaussian.

3.2.2 The Pivotal Choice

To address the non-normality in (5), two routes are possible, namely, one exploiting scale of fluctuations (modelling changing volatility), and the other exploiting shape of randomness (adding jumps to the continuous paths). This leads to the notion of 'states of randomness' introduced by Benoît Mandelbrot [81] to describe the level of smoothness of the price charts, that is, the irregularity due to jumps. He made a distinction between two types of states named 'mild randomness' and 'wild randomness'. He wrote that 'the roughness of a price chart [used to be] measured by its volatility—yet that volatility, analysts find, is itself volatile', and he described his contribution to be '[r]oughness is the very essence of many natural objects—and of economic ones' and to have developed a 'geometry of roughness'.

Models either work on the scale of fluctuations (the volatility part of risk) with unchanged shape of randomness (Brownian paradigm), or on shape of the randomness (the jump part of risk) with unchanged scale (constant volatility). In the first case, one moves outside the IID world because, as Houthakker [71] already observed in the 1960s, variance will change over time—volatility is itself volatile—even though this solution permits one to keep the paradigm of the Gaussian world. If one wants to stay in the IID framework, as in the second case, one has to change the randomness and leave the Brownian paradigm. This is summarized in the scheme below.

$$\text{Turning point} = \text{choice} = \begin{cases} \text{ROUTE 1} \quad \text{Wild} \quad \text{IID with roughness} \\ \\ \text{ROUTE 2} \quad \text{Mild} \begin{cases} \text{IID with jumps} \\ \text{non IID} \end{cases} \end{cases} \tag{6}$$

The situation at the end of the 1960s can be described in the Kuhn's analysis: with this pivotal choice, finance academics have 'before [them] a number of competing and incommensurable solutions to these problems, solutions that [they] must ultimately evaluate for [themselves]' [45, 165].

3.2.3 IID and Rough Fluctuations: Into the Wild

Leaving the mild randomness was in fact Mandelbrot's initial suggestion in 1962. He did not simultaneously attempt to address all stylized facts, but rather tried to find the simplest stochastic process to replace Brownian motion, while keeping time scaling of risk and the IID hypothesis in place. This was accomplished by α-stable motion, a subclass of the family of Lévy processes due to Paul Lévy. Equation (5)—or better, a version written here with squared errors to display variance, that is, $\varepsilon_k^2 = \sigma^2 \times u_k^2$—then becomes

$$\varepsilon_k^\alpha = \gamma^\alpha \times \ell_k^\alpha, \tag{7}$$

where $\ell_k \to \mathcal{L}_\alpha$ and $\mathcal{L}_\alpha$ is an α-stable distribution with α the characteristic exponent (intensity of jumps) and γ a scale parameter corresponding to the volatility in the α-stable world. In this model, however, the variance (second moment) is infinite (except when $\alpha = 2$, for then we find classical volatility). To put it differently, to obtain non-normality with the IID property, Mandelbrot introduced infinite variance.

Equation (7) means that market risk equals constant scale parameter times intensity of jumps. The corresponding scaling relationship (2) we recall below

$$X(t \times a) \stackrel{d}{=} X(t) \times a^{1/2}$$

becomes

$$X(t \times a) \stackrel{d}{=} X(t) \times a^{1/\alpha} \tag{8}$$

What Mandelbrot did was, in the end, nothing more than changing the exponent, and one might think such model tinkering would be greeted with little criticism. It was, however, the assumption of infinite variance that spurred a vehement controversy between Mandelbrot and the advocates of portfolio theory and option pricing models [42].

In fact, there was an overlapping between Gaussian distribution on returns and the capacity to manage a portfolio following Markowitz's 1952 approach, and to obtain a price on a given derivative instrument following Black-Scholes' 1973 model. Gaussian distribution or returns were of vital importance for these objectives. This is to say that one could not conceive in the 1970s an options market without validating at least implicitly the Bachelier-Osborne model. More fundamentally, the notion itself of calculation of option pricing had its initial intellectual roots in the Brownian representation. Hence on the contrary, the rejection of the Brownian representation led to an incapacity to evaluate, and so to hedge, the options whose volume was beginning to rise exponentially on the financial markets. From that time on, it became necessary, even vital for financial activity itself, that the Brownian representation be recognized as usable in first estimation to create models of stock market returns. These factors encourage the finance academics to sidestep econometrics subtleties that plagued statisticians. Thus the statistical-probability model in financial theory became embedded in an intuitive axiom of Brownian representation. The research habit here belongs to the second de Bruin's characterization: *intuition as data*.

Another research habit can be exhibited. No financial theory existed in the financial academic field which could include the infinite variance argued by Mandelbrot. In this case, it is possible to paraphrase Aumann [51]: while 'attractive as hypothesis [the infinite variance], there is little theory built on them'. Finance academics were excessively focussed on internal and technical problems (related to portfolio management and option pricing) instead of on empirically motivated issues. Thus the statistical-probability model in financial theory became embedded in financial technical problems. The research habit here is the third de Bruin's characterization: *introversion*.

It was the aim of statisticians to overcome the inadequacies of Brownian motion by tackling the issue of discontinuities without accepting infinite variance. We elaborate below on this.

3.2.4 Model Tinkering with Poissonian Jumps

At the turning point of the 1970s, Press [82] and Merton [66] put forward this approach; their idea was to keep the Brownian representation and to add a compound Poisson-normal (CPN) process to the diffusive Brownian component. *Jump-diffusion models* were born. As Merton observed,

> the total change in the stock price is posited to be the composition of two types of changes: diffusion and jumps. The natural prototype process for the continuous component of the stock price change is a Wiener process, so the prototype for the jump component is a 'Poisson-driven' process [66].

If we consider what effects this modelling approach has on the classic Brownian representation of return dynamics we encountered in Eq. (1), this transforms the earlier equation into:

$$X(t) = \underbrace{\mu t + \sigma W(t)}_{\text{Brownian motion}} + \underbrace{\sum_{i=1}^{N_t} Y(t)}_{\text{Jump part}}. \tag{9}$$

A component $\sum_{i=1}^{N_t} Y(t)$ is added, with a Poisson process $N(t)$ with jumps having a normally distributed size $Y(t)$ (a stochastic size). The underbraced glosses vividly illustrate model tinkering: in order to describe jumps, a CPN process is added to the model, but this is *perfectly consistent with the Gaussian intuition* (mild randomness): jump size has a normal distribution. Non-normality, in other words, is obtained by adding a jump component. To say it differently, we create 'wild' behaviour with 'mild' ingredients.

3.2.5 Model Tinkering with Time-Varying Volatility

We now proceed to examine the universe of non-IID processes with time-varying volatility. The initial relation (5)—written again with squared errors to display variance as $\varepsilon_k^2 = \sigma^2 \times u_k^2$ now becomes:

$$\varepsilon_k^2 = h_k \times u_k^2 \tag{10}$$

The constant variance σ^2 is replaced by a time-varying variance h_k of which the value depends on k. The time-varying variance is autoregressive and conditional in the sense that the variance h_k depends on past values of the squared errors ε_k^2, usually described using the term *heteroskedastic* (from the Greek words *heteros*, different/time-varying, and *skedastos*, dispersion/variance). These characteristics of modelling gave the name of this family: the Auto-Regressive Conditional Heteroskedasticity models (ARCH) introduced in 1982 by Nobel Prize winner Robert Engle [83]. The Eq. (10) is the explicit generating equation of an ARCH process. The functional form of h_k is crucial because it is meant to capture such phenomena as clustering of large shocks. This is brought out best by considering, rather heuristically, a model in which the conditional time-varying variance is formulated as a linear model on squared perturbations:

$$h_k = \alpha_0 + \alpha_1 \varepsilon_{k-1}^2 \tag{11}$$

The corresponding model, called ARCH (1), is given by the following equation, with glosses as above:

$$\varepsilon_k^2 = \underbrace{\left(\alpha_0 + \alpha_1\, \varepsilon_{k-1}^2\right)}_{\text{Gaussian distribution}} \times \underbrace{u_k^2}_{\text{Gaussian noise}} \tag{12}$$

If one calculates the kurtosis coefficient of the marginal distribution with an ARCH model, that is $K(\varepsilon)$, one finds that a value greater than 3, which is the normal value

(Gaussian distribution). The interpretation of this is that the ARCH (1) model has tails heavier than the Gaussian distribution. With the temporal dependence of h_k, the marginal distribution then appears to be leptokurtic even though the conditional distributions are still Gaussian. Like epicycles, one has created non-normality by embedding Gaussian distributions in a Gaussian distribution; or less impressionistically, we see that the ARCH principle rescales an underlying Gaussian noise by multiplying it by the conditional time-varying standard deviation (square root of the variance), which is a function of the past values.

As one may suspect, it may sometimes have been necessary to postulate an excessively large number of lags to capture the fine structure of dependence [84], implying the necessity of estimating a large number of parameters, leading to a high order ARCH process. To be parsimonious in mathematical terms and statistical estimations, the ARCH model was therefore generalized in Generalized ARCH (GARCH) models by Bollerslev [85]. This is a generalization that makes it possible to reduce the number of mathematical lags in the squared errors—and thus to reduce the computational burden. The GARCH model provides a parsimonious parameterization and is consistent with volatility clustering pattern. The heuristic intuition of the GARCH approach is easily glanced from

$$h_k = \underbrace{\alpha_0 + \alpha_1 \varepsilon_{k-1}^2}_{\text{ARCH(1)}} + \underbrace{\beta_1 h_{k-1}}_{\text{G(1)}}, \tag{13}$$

where h_{k-1} as previously. More generally, a GARCH(p, q) process is defined as

$$h_k = \underbrace{\alpha_0 + \alpha_1 \varepsilon_{k-1}^2 + \cdots + \alpha_p \varepsilon_{k-p}^2}_{\text{ARCH(p)}} + \underbrace{\beta_1 h_{k-1} + \cdots + \beta_q h_{k-q}}_{\text{G(q)}}. \tag{14}$$

We again see how the embeddedness of Gaussian building blocks made it possible to create non-normality without abandoning the Gaussian tools. Unconditionally, the GARCH process is homoskedastic with non-normality. Conditionally, the GARCH process is heteroskedastic with normality.

An interesting motivation for choosing ARCH models—instead of wild randomness—can be found in an article of Elie et al. [86, 97, our translation, our emphasis] in the following sentence: 'ARCH models have allowed us to largely address the fat tails problem, with *keeping a Gaussian framework, which is much more suitable* than that of stable distributions'. This statement illustrates the fourth de Bruin's characterization: *instrumentalism*. Actually, the market behaviour was compatible with a range of probabilistic models. But ARCH models were developed for purely instrumental reasons. Instead of revising the randomness, mathematical epicycles were added to the model to solve the conflict with the data.

3.3 Ten Years After...

Towards the end of the 1990s, Gouriéroux and Le Fol expressed a view than many seem to have held onto at the time:

> [T]he recent inflation of basic model varieties and terminology GARCH, IGARCH, EGARCH, TARCH, QTARCH, SWARCH, ACD-ARCH reveals that this approach appears to have reached its limits, cannot adequately answer to some questions, or does not make it possible to reproduce some stylized facts [87, our translation].

Particularly, it was noticed that the kurtosis coefficient implied by the ARCH and GARCH models tended to be far less than the sample kurtosis observed in empirical return series: the ARCH-GARCH ways of obtaining non-normality were, then, incompatible with returns empirically observed. The ARCH approach was able to generate excess kurtosis, but it did not go far enough, and hence a new move had to be made. As for NERP solutions, BORP solutions were criticized for failure to adequately describe the empirical data.

At this time, it was acknowledge by finance academics that jump-diffusion processes formed a particular subclass of Lévy processes. While the earlier separation of a continuous Brownian source of market movements and a Poisson source creating discontinuities was simple and convenient, it significantly limited modelling applications. But did the probabilistic representation of market fluctuations ultimately entail the use of the Brownian diffusive component? The diffusive part of probabilistic representations is needed for the modelling of the small movements only in the case of finite activity. Only with finite activity, the process required the addition of another component. In the 1990s, understanding was reached that infinite activity was possible. The usefulness of the diffusive component disappeared and a pure jump process seemed to be sufficient to represent the entire stock market phenomenon, that is, its bumpiness at all scales. The argument is well described in by 2002 as follows:

> The rationale usually given for describing asset returns as jump-diffusions is that diffusions capture frequent small moves, while jumps capture rare large moves. Given the ability of infinite activity jump processes to capture both frequent small moves and rare large moves, the question arises as to whether it is necessary to employ a diffusion component when modelling asset returns [69].

But that is another story.

References

1. O'Neill, O.: A Question of Trust: The BBC Reith Lectures 2002. Cambridge University Press (2002)
2. Jorion, P.: Risk management lessons from the credit crisis. Eur. Financ. Manage. **15**(5), 923–933 (2009). doi:10.1111/j.1468-036X.2009.00507.x
3. White, L.J.: Markets: the credit rating agencies. J. Econ. Perspect. **24**(2), 211–226 (2010)

4. Pixley, J.: New Perspectives on Emotions in Finance: The Sociology of Confidence. Routledge, Fear and Betrayal (2012)
5. de Bruin, B.: Ethics and the Global Financial Crisis. Cambridge University Press (2015)
6. Danielsson, J., Embrechts, P., Goodhart, C., Keating, C., Muennich, F., Renault, O., Shin, H.S.: An academic response to Basel II. Technical Report 130, London School of Economics (2001). Financial markets Group
7. Markopolos, H.: No One Would Listen: A True Financial Thriller. Wiley (2010)
8. Benmelech, E., Dlugosz, J.: The alchemy of CDO credit ratings. J. Monetary Econ. **56**(5), 617–634 (2009). doi:10.1016/j.jmoneco.2009.04.007
9. Le Courtois, O., Quittard-Pinon, F.: Risk-neutral and actual default probabilities with an endogenous bankruptcy jump-diffusion model. Asia-Pacific Finan. Markets. **13**(1), 11–39 (2007). doi:10.1007/s10690-007-9033-1
10. Pagano, M., Volpin, P.: Credit ratings failures and policy options. Econ. Policy **25**(62), 401–431 (2010). doi:10.1111/j.1468-0327.2010.00245.x
11. Sheik, A., Qiao, H.: Non-normality of Market Returns. Technical report, J.P. Morgan Asset Management (2009). Strategic Investment Advisory Group
12. Bachelier, L.: Théorie de la spéculation. Annales scientifiques de l'Ecole Normale Supérieure pp. 21–86 (1900)
13. Bernstein, P.: Capital Ideas: The Improbable Origins of Modern Wall Street. Wiley (2005)
14. Dimand, R.W., Veloce, W.: Alfred Cowles and Robert Rhea on the predictability of stock prices. J. Bus. Inquiry **9**(1), 56–64 (2010)
15. Preda, A.: Framing Finance: The Boundaries of Markets and Modern Capitalism. University of Chicago Press (2009)
16. Rhea, R.: The Dow Theory. Fraser Publishing Library, New York (1932)
17. Davis, M., Etheridge, A.: Louis Bachelier's Theory of Speculation: The Origins of Modern Finance. Princeton University Press, Princeton (2006)
18. Mackenzie, D.: An Engine, Not a Camera: How Financial Models Shape Markets. MIT Press, Cambridge, Mass. (2006). doi:10.1080/15265160902874361
19. Markowitz, H.: Portfolio selection. J. Financ. **7**(1), 77–91 (1952). doi:10.1111/j.1540-6261.1952.tb01525.x
20. Sharpe, W.F.: Capital asset prices: a theory of market equilibrium under conditions of risk. J. Finan. **XIX**(3), 425–442 (1964)
21. Fama, E.: The behavior of stock-market prices. J. Bus. **38**(1), 34 (1965). doi:10.1086/294743
22. Fama, E.: Efficient capital markets: a review. J. Finan. **25**(2), 383–417 (1970)
23. Mehrling, P.: Fischer Black and the Revolutionary Idea of Finance. Wiley, New York (2005)
24. Cont, R.: Empirical properties of asset returns: stylized facts and statistical issues. Quant. Finan. **1**, 223–236 (2001). doi:10.1088/1469-7688/1/2/304
25. Kendall, M.: The analysis of economic time-series-part I: prices. J. Royal Stat. Soc. Series A (General) **116**(1), 11–34 (1953). doi:10.2307/2980947
26. Bouchaud, J.P., Potters, M.: Theory of Financial Risk and Derivative Pricing: From Statistical Physics to Risk Management. Cambridge University Press, Cambridge (2000)
27. Boyarchenko, S.I., Levendorskii, S.Z.: Non-Gaussian Merton-Black-Scholes Theory. World Scientific (2002)
28. Cont, R., Tankov, P.: Financial Modelling with Jump Processes. Chapman and Hall (2004)
29. Eberlein, E., Keller, U.: Hyperbolic distributions in finance. Bernoulli **1**(3), 281–299 (1995). doi:10.2307/3318481
30. Embrechts, P., Kluppelberg, C., Mikosch, T.: Modelling Extremal Events for Insurance and Finance. Springer (1997)
31. Jondeau, E., Poon, S.H., Rockinger, M.: Financial Modeling Under Non-Gaussian Distributions. Springer, London (2007). doi:10.1007/978-1-84628-696-4
32. Kemp, M.: Extreme Events: Robust Portfolio Construction in the Presence of Fat Tails. Wiley Finance (2011)
33. Kotz, S., Kozubowski, T., Podgorski, K.: The Laplace Distribution and Generalizations. Birkhauser, Basel (2001). doi:10.1007/978-1-4612-0173-1

34. Rachev, S.T., Mittnik, S.: Stable Paretian Models in Finance. Wiley (2000)
35. Schoutens, W.: Lévy Processes in Finance: Pricing Financial Derivatives. Wiley Finance (2003)
36. de Bruin, B.: Reducible and nonsensical uses of game theory. Philos. Soc. Sci. **38**, 247–266 (2008). doi:10.1177/0048393108315557
37. de Bruin, B.: On the Narrow Epistemology of Game-Theoretic Agents. In: Majer, O., Pietarinen, A.V., Tulenheim, T. (eds.) Games: Unifying Logic, Language, and Philosophy, pp. 27–36. Springer, Dordrecht (2009)
38. de Bruin, B.: Overmathematisation in game theory: pitting the Nash Equilibrium Refinement Programme against the Epistemic Programme. Stud. Hist. Philos. Sci. Part A **40**(3), 290–300 (2009). doi:10.1016/j.shpsa.2009.06.005
39. de Bruin, B.: Explaining Games. Springer, Dordrecht (2010). doi:10.1007/978-1-4020-9906-9
40. Walter, C.: Une histoire du concept d'efficience sur les marchés financiers. Annales. Histoire, Sciences Sociales **51**(4), 873–905 (1996). doi:10.3406/ahess.1996.410892
41. Walter, C.: Aux origines de la mesure de performance des fonds d'investissement. Les travaux d'Alfred Cowles. Histoire & Mesure **14**(1), 163–197 (1999). doi:10.3406/hism.1999.1506
42. Walter, C.: La recherche de lois d'échelles sur les variations boursières. In: P. Abry, P. Goncalvés, J. Lévy Véhel (eds.) Lois d'échelle, fractales et ondelettes, pp. 553–588. Hermès, Paris (2002)
43. Walter, C.: 1900–2000: un siècle de processus de Lévy en finance. In: Bensimon, G. (ed.) Histoire des représentations du marché, pp. 553–588. Michel Houdiard, Paris (2005)
44. Walter, C.: Le phénomène leptokurtique sur les marchés financiers. Finance **23**(2), 15–68 (2002)
45. Kuhn, T.S.: The Structure of Scientific Revolutions. The University of Chicago Press, Chicago (1962)
46. von Neumann, J., Morgenstern, O.: Theory of Games and Economic Behavior. Princeton University Press, Princeton (1944)
47. Nash, J.: Non-cooperative games. Ann. Math. **54**(1), 286–295 (1951). doi:10.2307/1969529
48. Selten, R.: Reexamination of the perfectness concept for equilibrium points in extensive games. Int. J. Game Theor. **4**(1), 25–55 (1975). doi:10.1007/BF01766400
49. Myerson, R.B.: Refinements of the Nash equilibrium concept. Int. J. Game Theor. **7**(2), 73–80 (1978). doi:10.1007/BF01753236
50. Selten, R.: Spieltheoretische Behandlung eines Oligopolmodells mit Nachfrageträgheid: Teil I: Bestimmung des dynamische Preisgleichgewichts. Zeitschrift für die gesamte Staatswissenschaft **121**(2), 301–324 (1965)
51. Aumann, R.: What is Game Theory Trying to Accomplish? In: Arrow, K., Honkapohja, S. (eds.) Frontiers of Economics, pp. 28–76. Blackwell, Oxford (1987)
52. Camerer, C.F.: Behavioral Game Theory: Experiments in Strategic Interaction. Princeton University Press, Princeton (2003)
53. Samuelson, L.: Evolutionary Games and Equilibrium Selection. MIT Press, Cambridge, Mass (1997)
54. Goeree, J.K., Holt, C.A.: Stochastic game theory: for playing games, not just for doing theory. Proc. Nat. Acad. Sci. U.S.A. **96**(19), 10564–10567 (1999). doi:10.1073/pnas.96.19.10564
55. Heifetz, A.: Game Theory Interactive Strategies in Economics and Management. Cambridge University Press, Cambridge (2012)
56. Tadelis, S.: Game Theory: An Introduction. Princeton University Press, Princeton (2013)
57. Mill, J.S.: On the definition of political economy; and on the method of philosophical investigation in that science. London Westminster Rev. **26**, 1–29 (1836)
58. Myerson, R.B.: Game theory analysis of conflict. Cambridge, Mass (1991)
59. Simon, H.: Reason in Human Affairs. Stanford University Press, Stanford (1983)
60. Friedman, M.: The Methodology of Positive Economics. In: Essays in Positive Economics, pp. 3–43. Princeton University Press, Princeton (1953)

61. Osborne, M.: Brownian motion in the stock market. Operations research **07**(March-April), 145–173 (1959). doi:10.1287/opre.7.2.145
62. Jorion, P.: Value at Risk: The New Benchmark for Managing Financial Risk. McGraw-Hill, New York (2007)
63. Lo, A.W., MacKinlay, A.C.: Stock market prices do not follow random walks: evidence from a simple specification test. Rev. Financ. Stud. **1**(1), 41–66 (1988). doi:10.1093/rfs/1.1.41
64. Daníelsson, J., Zigrand, J.P.: On time-scaling of risk and the square-root-of-time rule. J. Bank. Finan. **30**(10), 2701–2713 (2006). doi:10.1016/j.jbankfin.2005.10.002
65. Goodhart, C.: The basel committee on banking supervision. A History of the Early Years 1974–1997. Cambridge University Press, Cambridge (2011)
66. Merton, R.C.: Option pricing when underlying stock returns are discontinuous. J. Finan. Econ. **3**(1–2), 125–144 (1976). doi:10.1016/0304-405X(76)90022-2
67. Cox, J.C., Ross, S.A.: The valuation of options for alternative stochastic processes. J. Finan. Econ. **3**(1–2), 145–166 (1976). doi:10.1016/0304-405X(76)90023-4
68. Ball, C.A., Torous, W.N.: On jumps in common stock prices and their impact on call option pricing. J. Finan. **40**(1), 155–173 (1985). doi:10.2307/2328053
69. Carr, P., Geman, H., Madan, D.B., Yor, M.: The fine structure of asset returns: an empirical investigation. J. Bus. **75**(2), 305–332 (2002). doi:10.1086/338705
70. Larson, A.B.: Measurement of a random process in futures prices. Stanford university. Food Res. Inst. Stud. **1**, 313–324 (1960)
71. Houthakker, H.S.: Systematic and random elements in short-term price movements. Am. Econ. Rev. **51**(2), 164–172 (1961)
72. Alexander, S.: Price movements in speculative markets: trends or random walks. Ind. Manag. Rev. **2**, 7–26 (1961)
73. Roberts, H.V.: Stock-market "patterns" and financial analysis: methodological suggestions. J Financ. **14**(1), 1–10 (1959). doi:10.1111/j.1540-6261.1959.tb00481.x
74. Cootner, P.H.: Common elements in futures markets for commodities and bonds. Am. Econ. Rev. **51**(2) (1961)
75. Mandelbrot, B.: Sur certains prix spéculatifs : faits empiriques et modèle basé sur les processus stables additifs non gaussiens de Paul Lévy. Comptes rendus hebdomadaires des séances de l'Académie des sciences **254**(3), 3968–3970 (1962). ark:/12148/bpt6k4002d
76. Mandelbrot, B.: The variation of certain speculative prices. J. Bus. **36**(4), 394–419 (1963)
77. Mandelbrot, B.: Nouveaux modèles de la variation des prix (cycles lents et changements instantanés). Cahiers du séminaire d'économétrie **9**, 53–66 (1966)
78. Mandelbrot, B.: Formes nouvelles du hasard dans les sciences. Économie appliquée **26**, 307–319 (1973)
79. Mandelbrot, B.: Le problème de la réalité des cycles lents et le syndrome de Joseph. Économie appliquée **26**, 349–365 (1973)
80. Mandelbrot, B.: Le syndrome de la variance infinie et ses rapports avec la discontinuité des prix. Économie appliquée **26**, 321–348 (1973)
81. Mandelbrot, B.: Fractals and scaling in finance. Discontinuity, Concentration, Risk. Selecta Volume E. Springer, New York (1997). doi:10.1007/978-1-4757-2763-0
82. Press, S.J.: A compound events model for security prices. J. Bus. **40**(3), 317–335 (1967)
83. Engle, R.F.: Autoregressive conditional heteroscedasticity with estimates of the variance of United Kingdom inflation. Econometrica **50**(4), 987–1007 (1982)
84. Engle, R.F.: Estimates of the variance of U.S. inflation based upon the ARCH model. J. Money Credit Bank. **15**(3), 286–301 (1983). doi:10.2307/1992480
85. Bollerslev, T.: Generalized autoregressive conditional heteroskedasticity. J. Econometrics **31**(3), 307–327 (1986). doi:10.1016/0304-4076(86)90063-1
86. Elie, L., El Karoui, N., Jeantheau, T., Jeanblanc, M.: Les modèles ARCH sur les cours de change. In: Proceedings of the 3rd AFIR Colloquium, pp. 91–101. Istituto Italiano degli Attuari, Italy (1993)
87. Gouriéroux, C., Fol, G.L.: Volatilités et mesures de risques. Journal de la Société de statistique de Paris **138**(4), 7–32 (1997)

Super-Exponential Bubbles and Expectations: Theory, Simulations and Empirics

Matthias Leiss

Abstract Transient super-exponentiality is a well-known statistical regularity of financial markets and generally associated with unsustainable growth and bubbles. We contribute to the understanding of super-exponential dynamics by assessing it from two new angles. First, we introduce an agent-based model of super-exponential bubbles on a risky asset market with fundamentalist and chartist traders. We show analytically and by simulations, that their mutual interactions repeatedly generate super-exponentially growing prices. Moreover, we combine our agent-based model with the methodology of log-period power law singularities (LPPLS) often used for bubble econometrics. We introduce a third type of trader who employs the LPPLS framework to detect the existence of a bubble and invests accordingly, trying to ride the bubble while it lasts and to quit before the subsequent crash. We find that the presence of LPPLS traders increases market volatility. In part two, we construct risk-neutral return probability distributions from S&P 500 option prices over the decade 2003–2013. The data strongly suggests increasing option-implied return expectations prior to the crash of 2008, which translates into transient super-exponential growth expectations. Furthermore, we find evidence for a regime-change from an abnormal pre-crisis period to a "new normal" post-crisis. Granger-causality tests indicate that the Federal Reserve policy played a significant role in this transition.

Materials

In this chapter, we discuss, summarize and expand previous own work in the area of super-exponentiality in finance. For further details we refer the reader to Kaizoji et al. [46] and Leiss et al. [50] as well as the references therein.

M. Leiss (✉)
ETH Zürich, Professorship of Computational Social Science,
Clausiusstrasse 50, 8092 Zürich, Switzerland
e-mail: mleiss@ethz.ch

M. Leiss
ETH Risk Center, Scheuchzerstrasse 7, 8092 Zürich, Switzerland

E. Ippoliti and P. Chen (eds.), *Methods and Finance*,
Studies in Applied Philosophy, Epistemology and Rational Ethics 34,
DOI 10.1007/978-3-319-49872-0_6

1 Introduction

Models of a financial market as a complex system of interacting investors have become popular, as they exhibit many empirically observed statistical regularities such as a fat-tailed distribution of returns and volatility clustering [54]. Although the existence of transient super-exponential growth is a well documented empirical fact for many financial markets, it has been largely neglected in this stream of literature [46]. However, a theoretical understanding of the conditions for transient faster-than-exponential growth is crucial, since this characteristic has been associated with the build-up of instabilities, sometimes termed "bubbles", that often end in market turbulences and crashes [69].

Over the last decades, financial markets have attracted many natural scientists, as these markets were shown to exhibit scaling laws that resemble those of large physical systems of interacting particles [55]. Earlier works involve Pareto [61], who found that a power law applies to the distribution of incomes, and Mandelbrot [56], who proposed that price changes on financial markets followed a Lévy stable distribution with power law tails of exponent less than 2. Later studies found the exponent to be 3, thus ensuring the existence of the variance [57]. The underlying idea that economic variables may be explained by the interactions among market participants appears to be in contradiction with the hypothesis of informationally efficient markets [21], according to which prices are only moved by the arrival and incorporation of new information. However, especially since the Global Financial Crisis of 2008, many scientists have emphasized the importance of complex interactions and positive nonlinear feedback for the understanding of markets, a paradigm that also goes by the name of reflexivity [76]. A recent study by Filimonov and Sornette [25] based on a self-excited conditional Poisson Hawkes model finds the prevalence of reflexivity on the Chicago Mercantile Exchange increased significantly from 30 % in 1998 to over 70 % in 2010.

Many interaction mechanisms studied in the literature involve psychological and social phenomena of humans [1]. One of the strongest psychological traits is imitation from others, which can be rational in situations of uncertainty and asymmetric information. For example, Hong et al. [38] found that a mutual fund manager is more likely to buy (or sell) a certain stock if other managers in the same city are buying (or selling) that stock. As they control for location, they can only explain the effect in terms of an epidemic model in which investors spread information and opinions about stocks by word of mouth. Sornette and Johansen [72] present a simple model that shows how herding due to social imitation may lead to a super-exponential growth in the price.

Technically, super-exponentiality in finance arises when the rate of return increases itself. To see this, let us define the infinitesimal return of an asset with price $p(t)$ as commonly done in mathematical finance (see e.g. [6, 28]):

$$\frac{\mathrm{d}p(t)}{p(t)} = r\,\mathrm{d}t + \sigma\,\mathrm{d}W(t), \tag{1}$$

where r, σ and $\mathrm{d}W(t)$ are the rate of return, the volatility and infinitesimal increments of a Brownian motion, respectively. For simplicity we may consider only the deterministic equation, as everything carries over to the stochastic case. Thus

$$\mathbb{E}\left[\frac{\mathrm{d}p(t)}{p(t)}\right] = r\,\mathrm{d}t. \tag{2}$$

Usually, the rate of return r is left constant, but let us assume it grows linearly in time:

$$r(t) = r_0 + \gamma t, \tag{3}$$

with r_0 and γ constants. Then the solution of Eq. (2) is given by

$$\mathbb{E}\left[p(t)\right] = p_0\,\mathrm{e}^{r_0 t + \gamma t^2/2}. \tag{4}$$

For $\gamma = 0$ we recover the well-known standard exponential growth due to compounding interests, which is commonly found in economic processes and reflects Gibrat's law of proportional growth [30]. However, for positive $\gamma > 0$ the rate of return itself grows in time such that the price increases much faster than an exponential, which we refer to as "super-exponential" growth. Of course, for rates of return that exhibit a stronger than linear transient growth dynamics, we can expect an even faster increase in the price of the asset. In general, such a growth path is not sustainable and therefore of a transient nature. Thus, it is usually associated with the build-up of instabilities, that in finance are often termed bubbles.

Empirically, super-exponential growth has been found on many financial markets, e.g. for stocks and equities [42, 78], oil [75], real estate [79, 81] as well as commodities and derivatives [73]. Furthermore, Hüsler et al. [39] recently showed via controlled experiments in the laboratory, that a combination of over-optimistic expectations with positive feedback lead to prices that grow particularly fast as $p(t) \sim \mathrm{e}^{\mathrm{e}^t}$. Sometimes super-exponential growth gives a sufficiently strong signal for ex-ante bubble prediction [44, 51, 70, 71].

Theoretically, there exist at least two frameworks allowing for super-exponentiality. The first is an extension of the rational expectations model by Blanchard [7]; Blanchard and Watson [8] based on discrete scale invariance by Johansen et al. [43, 44]. In this so-called JLS model, crashes occur randomly at a certain hazard rate. It is rational for traders to stay invested as long as they are compensated by a higher rate of return accounting for the risk of a crash. However, this establishes a proportionality condition between the hazard rate and the expected rate of return. As one increases the other follows, until a crash stops the unsustainable vicious circle. A remarkable feature of the JLS model are specific predictions about the presence of certain price patterns prior to the crash, so-called log-periodic power law singularities (LPPLS) that arise from hierarchical positive feedback among traders [72, 74].

The second theoretical framework is the one of (mildly) explosive bubbles by Phillips et al. [63, 64]. Describing the log-price as an autoregressive process with coefficient slightly larger than, but over time decreasing to one implies super-exponential growth of the price. This inspired empirical bubble tests based on Markov-switching state-space models alternating between stable random walks and unstable explosive regimes [2, 49]. The tests consist in detecting structural breaks indicating the transition from one regime to the other by using, for example, Chow-type procedures [15]. Both directions have been investigated, i.e. the onset of bubble [37, 63, 64] and its end [9].

Agent-based models (ABM) represent a natural way to study the aggregate outcome due to interactions among possibly heterogeneous individuals. In a first important contribution, De Long et al. [16, 17] employed an ABM to quantitatively study a financial market populated by "rational" and "noise" traders [5, 48]. While rational traders base their investment decision on fundamental values, noise traders exhibit erroneous stochastic beliefs leading to an unpredictable additional risk in the asset price. As a result, rational investors fail to exploit the noise traders' irrational behavior and market prices significantly deviate from fundamentals. A number of works have extended the set-up of noise traders to account for group psychological and sociological effects such as herding and trend-following [10–14, 47, 54, 55].

Although successful in explaining many statistical regularities of financial markets such as a fat-tail distribution of returns and volatility clustering, to our best knowledge no ABM has studied the link between interactions among investors and transiently super-exponentially growth of asset prices. However, such an approach is crucial for a qualitative and quantitative micro-understanding of super-exponentiality. This gap in the literature has been closed by Kaizoji et al. [46], who starting from well-known set-ups of agent-based models for financial markets, analytically and in simulations derive conditions for these explosive price paths. In Sect. 2 we will summarize their results and extend the model by including a third group of investors, who try to arbitrage the arising super-exponential price patterns based on the LPPLS methodology by Sornette and Johansen [72]; Sornette et al. [74]; Filimonov and Sornette [26]. Philipp [62] quantifies the impact of LPPLS traders on the market as a whole and on the development of bubbles in particular. The presence of LPPLS investors is found to increase a bubble's peak proportional to their market power, but not its duration.

The second part of this document is dedicated to empirical observations of investors' return expectations. Following Leiss et al. [50], we estimate risk-neutral probability distributions from financial option quotes on the S&P 500 stock index over the period 2003–2013. The employed method by Figlewski [23] is an essentially model-free technique, allowing for nonstandard density features such as bimodality, fat tails and general asymmetry, and thus is particularly suited to study the profound impacts of the Global Financial Crisis of 2008. Evaluating the resulting risk-neutral distributions in terms of their moments, tail characteristics and implied returns reveals three different regimes: a pre-crisis, crisis and post-crisis phase. Interestingly, the pre-crisis period is characterized by linearly rising returns as in Eq. (3), which translates into super-exponential growth expectations of the representative investor under risk neutrality. Granger-causality tests show that expected returns lead yields on 3-month Treasury Bills prior to the crisis, while the inverse is true in the post-crisis period.

2 An Agent-Based Model of Super-Exponential Bubbles

Many studies have employed agent-based models to explain the statistical regularities of financial markets such as a fat-tail distribution of returns and volatility clustering [54, 55, 57]. However, the empirically well-established fact that instabilities and bubbles can be related to super-exponential growth of asset prices was, to our best knowledge, neglected. Kaizoji et al. [46] set out to show, both theoretically and in simulations, that super-exponential growth is nested in the widespread agent-based models. After setting up the model, we theoretically derive conditions for faster-than-exponential growth in an asset price, which will be tested in numerical simulations. Finally, we extend the model by adding investors to the market who try to detect and exploit super-exponential patterns.

2.1 Set-Up of the Model

Our model describes a financial market with a dividend-paying risky asset in fixed supply and a risk-free asset in unlimited supply. The price of the risky asset is determined by market clearing at each time step, whereas the risk-free asset guarantees a known and constant rate of return. There are two types of investors on the market, who dynamically allocate their wealth between the risky and the risk-free asset. We follow popular contributions to describe their respective investment strategies. On the one hand, fundamentalists exhibiting a constant relative risk aversion (CRRA) balance return and risk of the risky asset in order to maximize the expected utility of next period's wealth [12–14]. The dividend-price ratio is modeled as an i.i.d. random variable with stationary normal distribution. Since fundamentalist investors know and understand this process, they decide to invest a constant fraction of their wealth in the risky asset.

On the other hand, chartist traders perform technical analyses (momentum trading) and are subject to social influence [54, 55]. The technical analysis consists of detecting and following trends in the risky-asset price by employing an exponentially weighted moving average of past returns. Social influence involves a tendency to imitate the majority opinion among chartists. Individually chartist traders are polarized in their investment decision, i.e. go fully in or out the risky asset. They switch from one state to the other by weighing both momentum and social influence on equal footing. Furthermore, we allow for idiosyncratic noise. Note, that due to our stochastic framework and the large number of investors their *aggregate* impact on the market is continuous. Mathematically the setup of chartist traders equals the description of a spin system such as the Ising model subject to an external magnetic field [40].

For our analysis here it is sufficient to simplify the full model by only studying the deterministic set of equations, where random variables are replaced by their mean. For a derivation and the complete set of stochastic equations we refer to Kaizoji et al. [46].

Dynamics of the chartists opinion index s:

$$s_t = (1 + \kappa - p)s_{t-1} + \kappa H_{t-1}\,, \tag{5}$$

where κ and p measure the importance of trend following and social imitation, as well as idiosyncratic decisions, respectively.

Dynamics of the risky asset price P:

$$\frac{P_t}{P_{t-1}} = \frac{\sum_{j=f,c} x^j_t W^j_{t-1}\left[R_f(1 - x^j_{t-1}) + r x^j_{t-1}\right]}{\sum_{j=f,c} x^j_{t-1} x^j_t W^j_{t-1}}, \tag{6}$$

where x^f and $x^c = (1 + s)/2$ are the fractions of wealth invested in the risky asset by fundamentalists and chartists, respectively, R_f is the gross rate of return of the risk-free asset and r the average dividend-price ratio.

Wealth of fundamentalists W^f:

$$W^f_t / W^f_{t-1} = x\left(\frac{P_t}{P_{t-1}} + r\right) + (1 - x)R_f\,. \tag{7}$$

Wealth of chartists W^c:

$$W^c_t / W^c_{t-1} = \frac{1 + s_{t-1}}{2}\left(\frac{P_t}{P_{t-1}} + r\right) + \frac{1 - s_{t-1}}{2} R_f\,. \tag{8}$$

Momentum of the risky asset price H:

$$H_t = \theta H_{t-1} + (1 - \theta)\left(\frac{P_t}{P_{t-1}} - 1\right), \tag{9}$$

where θ describes the timescale for the chartists' technical analysis. Note, that the price and wealth level equations (6–8) suggest in their formulation the nature of multiplicative growth, which underlies so many economic and financial systems [65].

2.2 Theoretical Analysis

Assuming slowly varying relative wealth differences between the two investor groups allows us to decouple the wealth equations from the others. In this case, a stability analysis shows that there exists one fixed point (s^*, H^*). It is stable in the regime $\kappa < p$, i.e. when idiosyncrasies overweigh the ordering forces of herding and trend following, and unstable otherwise. Starting with an initial opinion index s_0, the pricing equation one may approximate the pricing equation (6) as

$$\frac{P_t}{P_{t-1}} = 1 + b(s_t - s_{t-1}) + \mathcal{O}(r, R_f, (s - s_0)^2) \,, \tag{10}$$

where $b \sim \mathcal{O}(1)$ is a constant of order one. Therefore, up to the same level of approximation, the price transiently changes as

$$\frac{P_t}{P_0} = \prod_{j=1}^{t} \left[1 + b(s_j - s_{j-1})\right] \simeq \prod_{j=1}^{t} \mathrm{e}^{b(s_j - s_{j-1})} = \mathrm{e}^{b(s_t - s_0)} \,, \tag{11}$$

i.e. exponentially in the opinion index s. But as Eq. (5) shows, in the unstable regime $\kappa > p$ the opinion index grows exponentially in time (until it saturates). This means that the price transiently grows as the exponential of an exponential in time, or super-exponentially.

The prevalence of herding and trend-following arguably varies over time, as the socio-economic context as well as financial markets themselves change. The economics literature refers to this effect as *regime switching* (see for example [33, 34]). We follow Harras et al. [35], propose that κ is time-dependent and model it as a discretized mean-reverting Ornstein-Uhlenbeck process with mean in the stable regime $\kappa < p$.

$$\kappa_t - \kappa_{t-1} = \eta(\mu_\kappa - \kappa_{t-1}) + \sigma_\kappa \nu_t \,, \tag{12}$$

where $\eta > 0$ is the mean reversion rate, μ_κ is the mean reversion level and $\sigma_\kappa > 0$ is the step size of the Wiener process realized by the series ν_t of standard i.i.d. standard normal random variables $\sim N(0, 1)$. This is related to Lux [53], who enforces regime switching between bearish and bullish markets with a deterministic term proportional to returns in the transition probabilities, while our approach adds stochasticity to the strength of social imitation and trend following.

2.3 Simulation Results

The simulation parameters are calibrated to empirical values, if possible. For example the daily risk-free rate of interest compounds to an annualized rate of 2 %, and the dividend-price ratio is taken from Engsted and Pedersen [20].

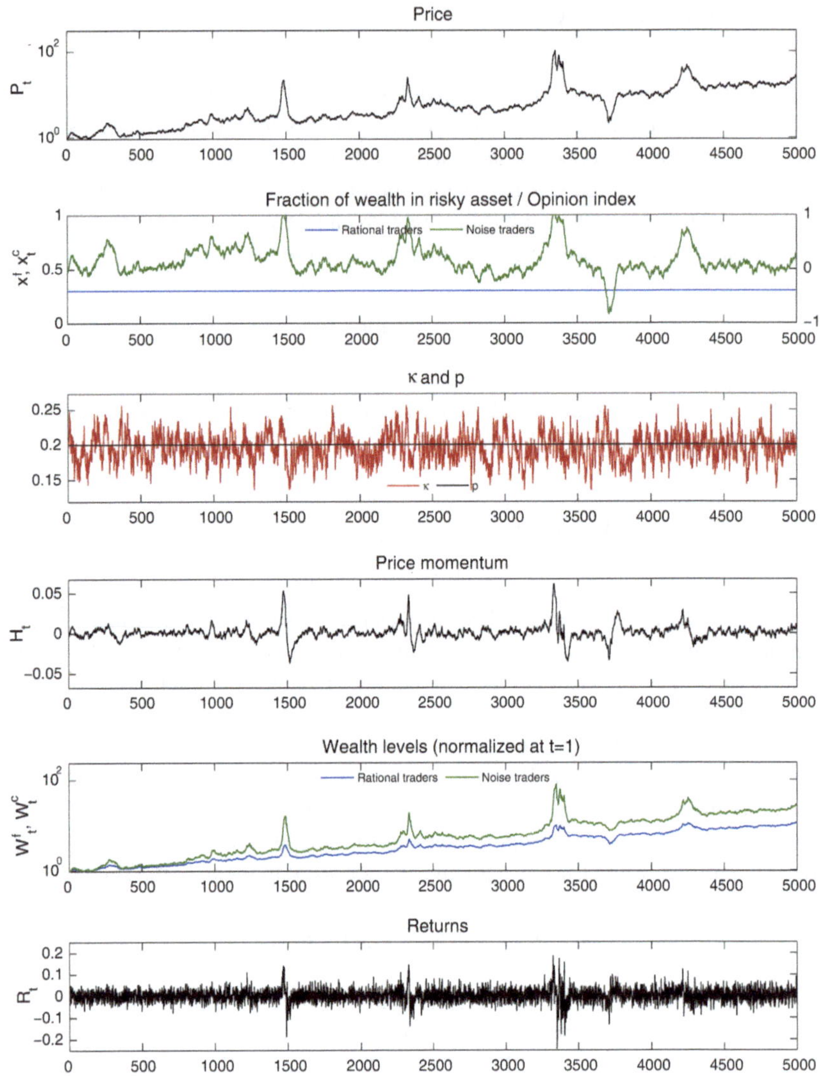

Fig. 1 Time-series generated by a representative run of the agent-based model of [46]

$$\theta = 0.95, \quad r = 1.6 \times 10^{-4}, \quad \sigma_r = 9.5 \times 10^{-4}, \quad R_f = 8 \times 10^{-5}, \quad x^c = 0.3\,, \tag{13}$$

$$p = 0.2, \quad \mu_\kappa = 0.196, \quad \sigma_\kappa = 0.001, \quad \eta = 0.11\,. \tag{14}$$

Figure 1 shows the time dependence of the relevant variables P_t, x_t^c, κ_t, H_t, W_t^f, W_t^c and the daily returns R_t during a representative run of the model when both investor groups start with equal endowments. The price P_t has a general upward trend driven by an inflow of money into the system due to the risk-free rate of return and dividends subject to noise. However, one may clearly identify five phases of extreme price movements, i.e. bubbles, out of which four are positive and one actually is a negative bubble.[1] Figure 2 shows one of these bubbles in detail. The growth of the logarithm of the price itself transiently accelerates, i.e. the price grows super-exponentially in time as explained by Eq. (11). This behavior is driven by the an increasing level of social imitation among chartists indicated by a polarizing opinion index. It is interesting to see that the super-exponential growth of the price is modulated by oscillations which resemble the trajectory of a log-periodic power law singularity [72, 74]. Figure 3 shows the distribution of returns as well as the autocorrelation of signed and absolute returns. As extensively documented in the literature, the statistical regularities of such an agent-based model consist of a fat-tail distribution of returns, a fast-decaying autocorrelation function of signed returns and a slowly decaying autocorrelation function of absolute returns equal those empirically observed on financial markets [54].

2.4 The Effect on LPPLS Traders on the Market

Since the bubbles generated by the model defined in Sect. 2.1 exhibit super-exponential growth behavior modulated by approximate log-periodic oscillations, it is intriguing to enrich the model with a third group of investors who employ try to detect and exploit them based on the LPPLS methodology [72, 74]. How will they impact on the market price in general and especially during bubbles? The efficient market hypothesis claims the price of an asset to be near its fundamental value as judged by the market. Investors with superior information of temporary deviations from fundamentals will use it for arbitrage and push prices back to fundamentals, leading to full efficiency. Already Friedman [29] argued that this mechanism would be even more effective in the long run, since agents with persistently superior knowledge would survive, while the others would lose their capital and eventually be driven out of the market.

Reflexivity, however, can lead to market outcomes that differ strongly from what the efficient market hypothesis predicts. In his conventionalist approach André Orléan argued that prices were essentially the result of interactions driven by the various beliefs of market participants Orléan [59]. It could then be possible that prices stabilize due to self-referential interactions at levels different from the fundamental values as predicted by the EMH—this is what Orléan calls a "*convention*". But then the dominance of investors with superior information of fundamentals will no longer be guaranteed because "*markets can remain irrational a lot longer than you and I can remain solvent.*" (attributed to John Maynard Keynes, see [68]).

[1] Yan et al. [78] define a negative bubble as the "mirror image of standard financial bubbles, in which positive feedback mechanisms may lead to transient accelerating price falls.".

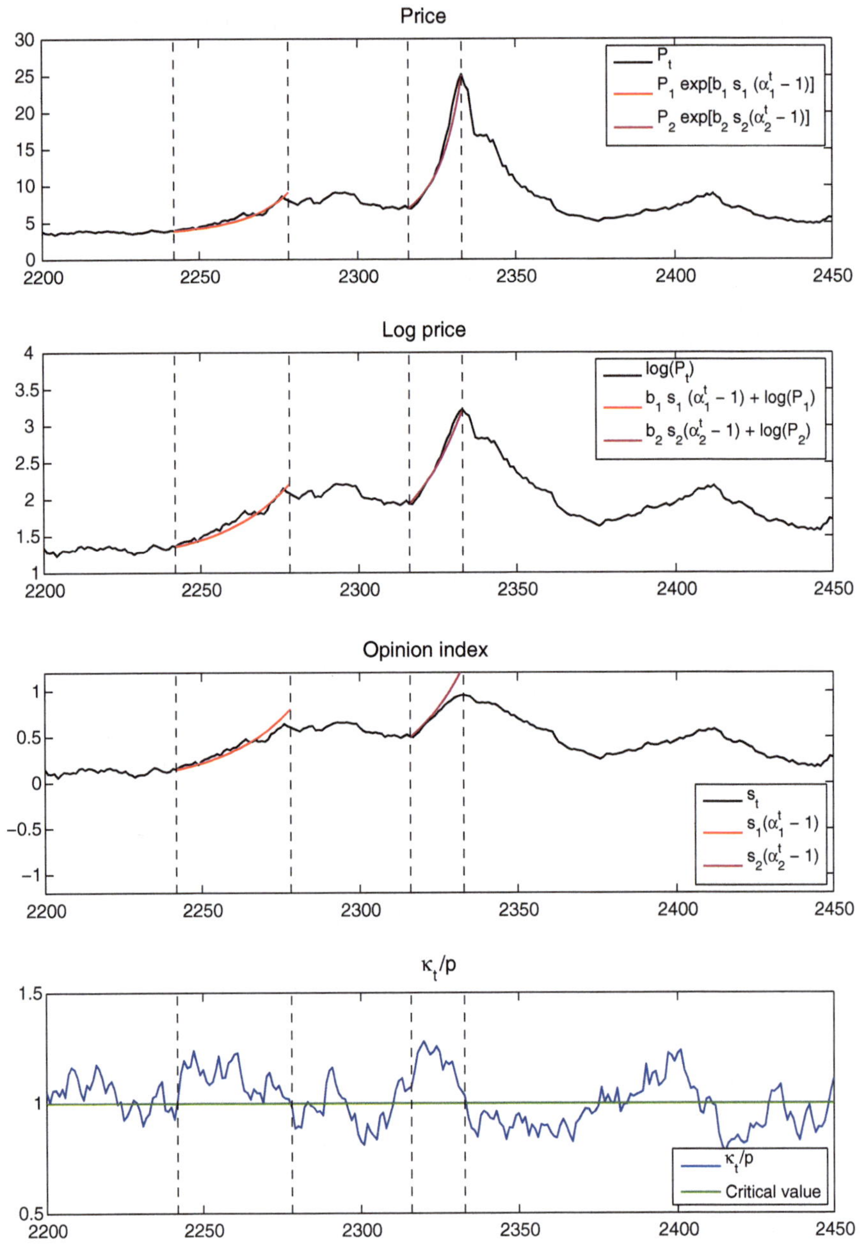

Fig. 2 Close-up view on one of the bubbles shown in Fig. 1, where s_i, α_i, b_i for $i = 1, 2$ are constants [46]

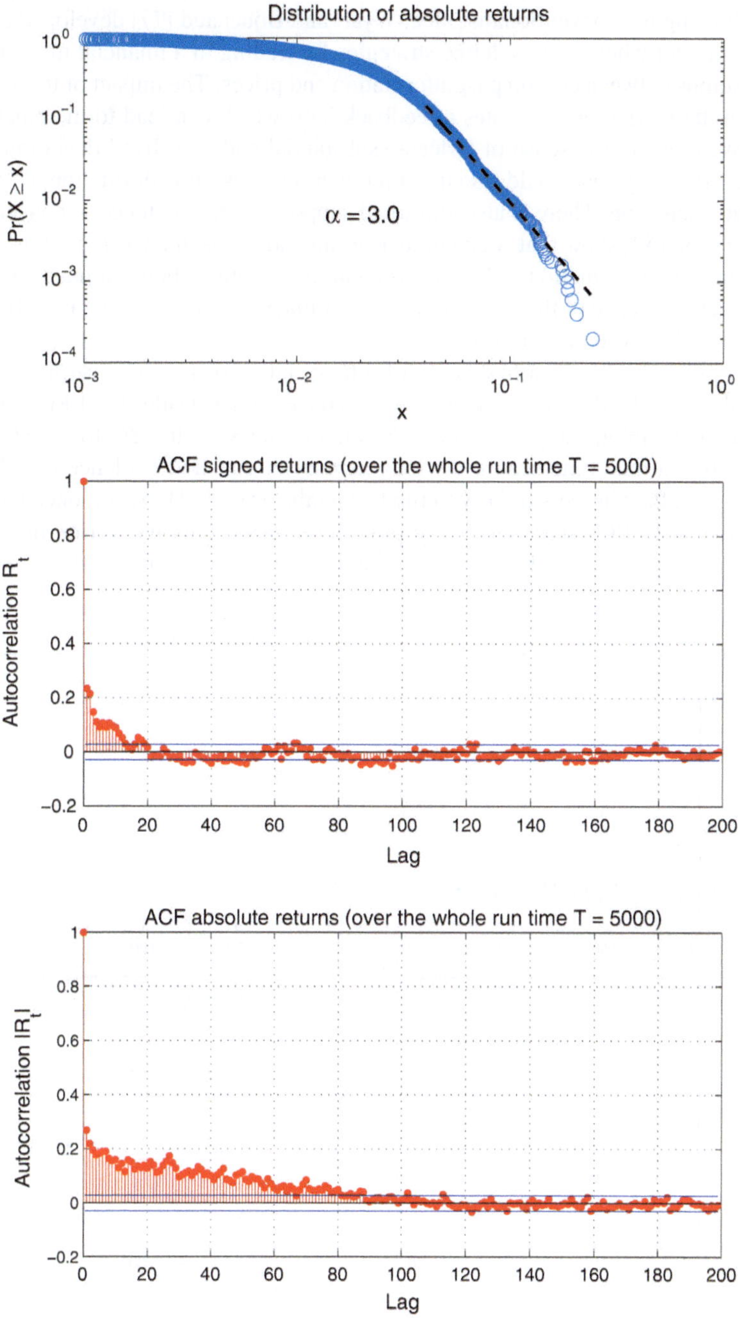

Fig. 3 Distribution of absolute returns, autocorrelation of signed and absolute returns of the run presented in Fig. 1 [46]

Following the conventionalist idea, Wyart and Bouchaud [77] developed a quantitative model where agents define strategies for trading in a financial market using correlations between certain past information and prices. The impact of these strategies on the market price creates a feedback loop which can lead to the emergence of conventions in the sense of Orléan—substantial and long-lived deviations from market efficiency. One could also interpret their result as a transformation of correlation into causation. There is also empirical support for the existence of conventions. Lorenz et al. [52] show that social influence can undermine the wisdom of the crowd effect in simple estimation tasks. In particular, information about estimates of others led to a convergence of the group's average estimate that often is farer away from the true value than when no information is given.

Philipp [62] presents a first study of a financial market as in Kaizoji et al. [46] with additional LPPLS trades in favor of the conventionalist side. The LPPLS traders use the methodological framework by Filimonov and Sornette [26] to detect bubble signals on various time scales. As long as they do not find evidence of a bubble building up, they invest similarly to fundamentalist traders. However, once they find a significant LPPLS signature in the price time series, they try to ride the bubble by fully investing until shortly before the anticipated crash. Figure 4 presents the

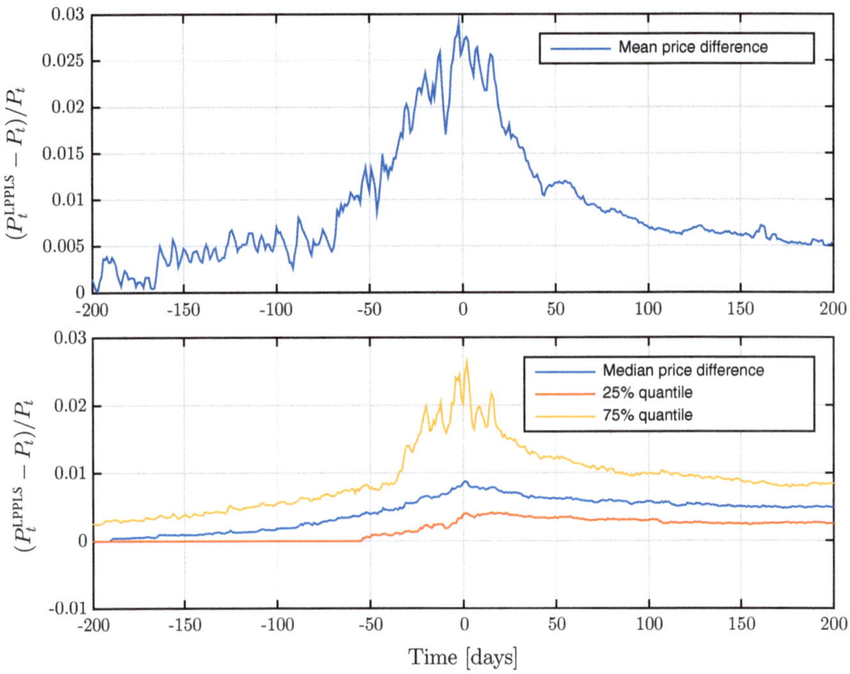

Fig. 4 The mean and median effect of LPPLS traders with 3 % wealth share on the risky asset price during a bubble as compared to a market without LPPLS traders. Reprinted with permission from [62]

average and median effect of LPPLS traders on the price during a bubble. Philipp [62] finds the presence of LPPLS investors to increase a bubble's peak proportional to their market power, but not its duration.

3 Evidence from the Options Market

In the previous section we set out to build a model that explains market turbulences via unsustainable growth expectations of chartist traders driven by social imitation and trend following. In this section, following Leiss et al. [50], we will infer return expectations by computing risk-neutral probability distributions from financial option prices over the decade around the Global Financial Crisis of 2008. The resulting distributions show clear regime shifts, which allows us to endogenously separate the decade 2003–2013 into a pre-crisis, crisis and a post-crisis regime. Very interestingly, we find super-exponential growth expectations prior to the Global Financial Crisis until mid-2007, and a 'new normal' regime since 2009. This validates the kind of price patterns simulated and discussed in the previous section. Finally, we determine the direction of Granger causality between expected returns and 3-month Treasury Bill yields for the sub-periods, respectively.

3.1 Estimation of Risk-Neutral Densities

Simply speaking a financial option is a bet on the future price of the underlying security. Since options are traded on public exchanges, the price of the option reflects the probability distribution of outcomes corresponding to the bet as seen by the market. Mathematically, this is captured by the fundamental theorem of asset pricing stating that the price of a security is given by the discounted payoff expected under risk neutrality [18]. Denoting the risk-neutral measure $\mathbb{Q}$ and the risk-neutral density by f, respectively, today's price C_0 of a standard European call option with exercise price K and maturity T on a stock with price at maturity S_T is given by

$$C_0(K) = \mathrm{e}^{-r_f T}\, \mathbb{E}_0^{\mathbb{Q}}\left[\max(S_T - K, 0)\right] = \mathrm{e}^{-r_f T} \int_K^{\infty} (S_T - K) f(S_T) \mathrm{d}S_T, \tag{15}$$

where r_f is the risk-free rate of return. In expression (15) all quantities but the risk-neutral density are observable. This allows us to invert the equation and numerically determine the density $f(S_T)$, which describes the representative investor's expectations for the future price under risk-neutrality.

Jackwerth [41] gives an excellent overview over several methods to numerically construct a risk-neutral density from option prices according to (15). We will follow a non-parametric method proposed by Figlewski [23] that combines standard smoothing techniques in the implied-volatility space with generalized extreme value

(GEV) density fitting for the tails.[2] The essentially model-free approach will turn out to be necessary to account for strongly changing contexts in the course of the Global Financial Crisis of 2008. It is able to reveal non-standard features of risk-neutral densities such as deviations from log-normality, tail asymmetries and bimodality.

For completeness we will briefly review the method here and refer the reader for more detail to Leiss et al. [50] and Figlewski [23]. We start with bid and ask quotes for standard European call and put options on the S&P 500 stock index for a certain maturity, respectively. As a first step, we filter out very deep out-of-the-money options, as their spreads are as large as the option midprice implying huge uncertainty. Next, we may combine call and put midprices and only use the more liquid at-the-money and out-of-the-money data. Using the inverted Black-Scholes model we calculate implied volatilities and smoothly blend them from puts to calls in the at-the-money region [6]. A fourth-order polynomial is fitted in implied-volatility space and via the Black-Scholes equation transformed back to price space.[3] Differentiating equation (15) with respect to the strike price gives us the risk-neutral distribution and density, respectively:

$$F(S_T) = \mathrm{e}^{r_f T} \frac{\partial}{\partial K} C_0(K) + 1, \quad f(S_T) = \mathrm{e}^{r_f T} \frac{\partial^2}{\partial K^2} C_0(K). \tag{16}$$

Thus, we take the numerical second derivative of the fit in price space to construct the risk-neutral density over the range of observable and used strike prices. Beyond this range we append tails of the GEV family to the left and right, respectively. Both are characterized by three parameters each, that are determined by matching the density bulk at two quantile points and the conservation of probability mass. Figure 5 illustrates the method.

Following Leiss et al. [50], we use end-of-day bid and ask data of call and put options written on the Standard & Poors 500 stock index over the time period of 2003–2013. We limit our analysis to maturities in March, June, September and December, which are the most liquid options. We exclude the 14-day window before expiration, as the shrinking range of relevant strikes only leads to highly peaked densities. Midprices for which no implied volatility exists are discarded, too. This leaves us with 2,311 observations.

3.2 Moment Analysis and Regime Changes

In this section, we will analyze the time dynamics of the estimated risk-neutral densities over the whole time period. We will characterize a risk-neutral density estimated

---

[2]The family of generalized extreme value distributions contains the limit distribution of adequately normalized maxima of a sequence of i.i.d. random variables, provided the limit exists. It is used extensively in finance and risk management practice [19].

[3]By contrast to Figlewski [23], Leiss et al. [50] use open interest to weigh the fit. This should increase the importance of those data points which reflect more market information.

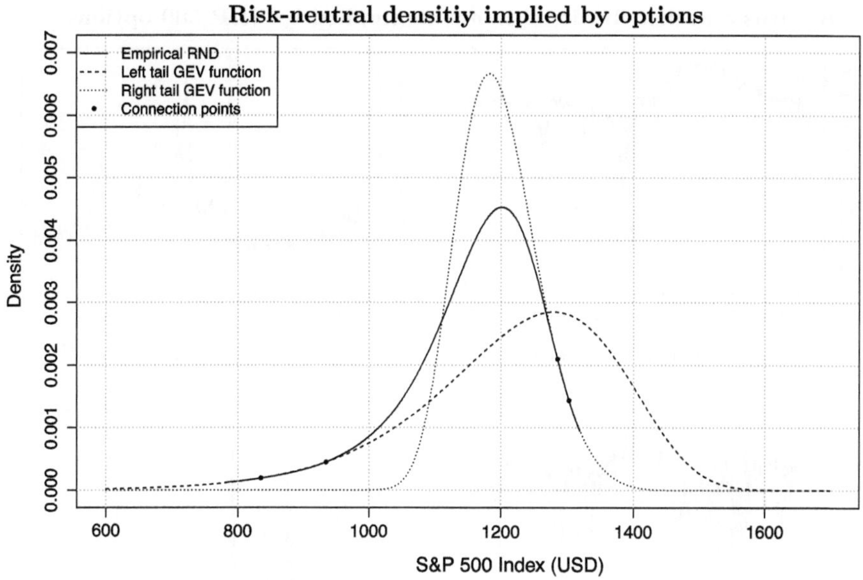

Fig. 5 Risk-neutral density implied by S&P 500 options from 2010-10-06 for index levels on 2010-12-18 [50]

on a trading day by the second to fourth moment as well as the shape parameters of the left and right GEV tail, respectively.[4] Furthermore, we determine the return r_t implied by options expected under risk-neutrality as:

$$r_t^{\mathrm{Q}} = \frac{1}{T-t} \int_0^{\infty} \log\left(\frac{S_T}{S_t}\right) f(S_T) \mathrm{d}S_T, \tag{17}$$

where S_t is the price of the underlying at the estimation time t. As can be seen in Fig. 6, the time series clearly undergo regime changes in the course of the Global Financial Crisis of 2008. Leiss et al. [50] formally analyze the option-implied returns, moments, tail-shape parameters for change points based on the binary segmentation algorithm with the cumulative sum test statistic [60, 67]. The results are presented in Table 1 and show that based on the left tail-shape parameter and the second to fourth moment one may consistently define the crisis period from mid-2007 to 2009. For the following analysis we take the dates suggested by changes in the left-tail shape parameter and define the crisis period as June 22nd, 2007 to May 4th, 2009, which is in line with the collected occurrence of events by the [22]. It is interesting to note that when looking at expected returns, the onset of the crisis manifested itself only more than a year later, on September 5th, 2008.

[4]By construction, the first moment equals the discounted price of the underlying at the time of estimation. As it does not carry new information, we will omit it in our analysis.

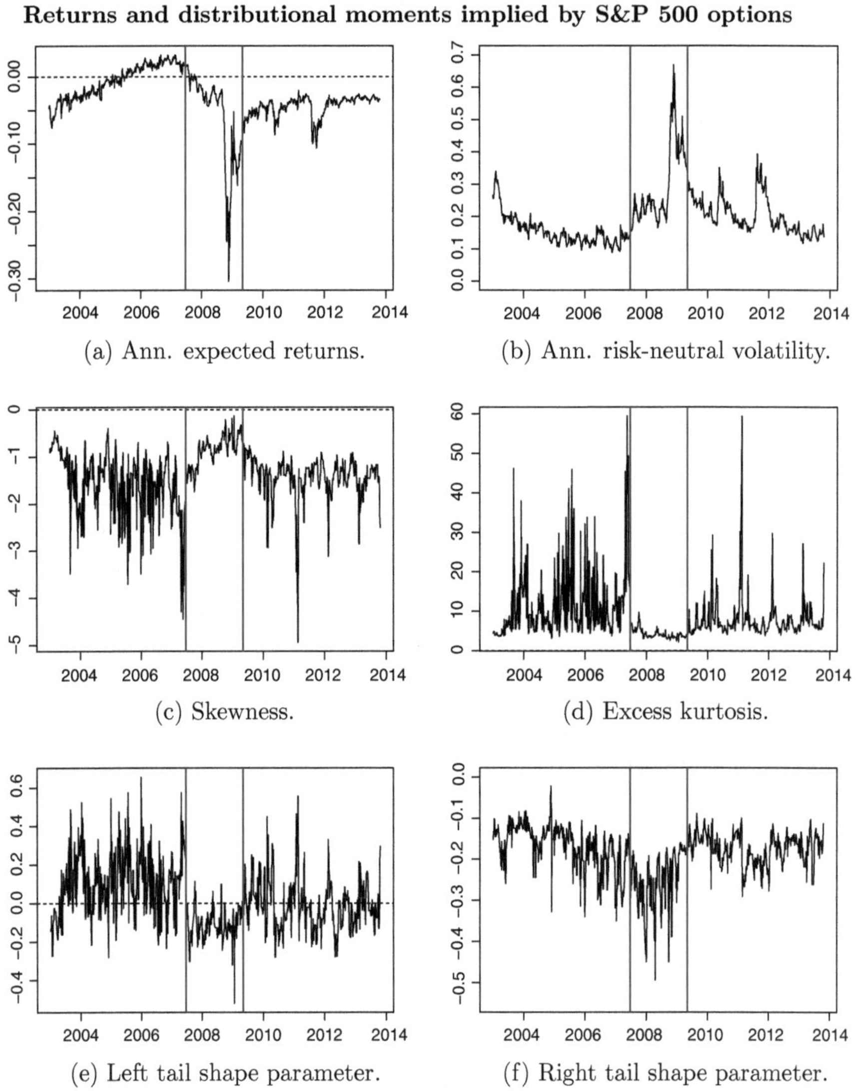

Fig. 6 This figure shows returns and distributional moments of the risk-neutral densities as implied by S&P 500 options. *Red vertical lines* indicate the endogenously identified period of the Global Financial Crisis [50]

Over the course of the decade 2003–2013, both the crisis and the presence of risk-aversion led to option-implied returns, which were negative on average with a mean value of −3 % with a standard deviation of 4 %. The normalized annualized second moment of the price densities is called risk-neutral volatility [24].[5] Over the whole

[5]Normalization means to evaluate $f(S_T/S_t)$ instead of $f(S_T)$.

Table 1 Start and end dates of the Global Financial Crisis as identified by a change point analysis of statistical properties of option-implied risk-neutral densities [50]

Variable	Crisis start date	Crisis end date
Left tail shape parameter	2007-06-22***	2009-05-04***
Right tail shape parameter	2005-08-08***	2009-01-22***
Risk-neutral volatility	2007-07-30***	2009-11-12***
Skewness	2007-06-22***	2009-10-19***
Kurtosis	2007-06-19***	*NA*[a]
Option-implied returns	2008-09-05***	2009-07-17***

Note *p < 0.1; **p < 0.01; ***p < 0.001
[a]No change point indicating a crisis end date found

period, it averages to 20 % (standard deviation of 8 %), and increases during the crisis period to 29 ± 12 %. Although exhibiting large fluctuations, skewness (-1.5 ± 0.9) and excess kurtosis (10 ± 12) strongly deviate from log-normality, as assumed for example by the Black-Scholes option pricing model [6]. Curiously, but in line with [4], we observe a skewness (-0.9 ± 0.3) and excess kurtosis (4.4 ± 1.6) more similar to those of a log-normal distribution during the crisis period. One interpretation of this result is that during a crisis investors focus more on the immediate exposure to the market, while neglecting tail risks indicating future extreme events. Another, more technical explanation that involves conditional estimations is given in [50]. The tail shape parameter of a GEV distribution determines the likelihood of extreme events. Over the whole period, a shape parameter of -0.19 ± 0.07 for the right tail means that the representative investor does not expect big upward jumps under risk-neutrality. By contrast, a left tail shape parameter of 0.03 ± 0.23 indicates a thin distribution of large losses, albeit with a surprisingly small probability.

3.3 *Super-Exponential Growth Prior to the Global Financial Crisis*

One of the important features of the time series of both option-implied and realized S&P 500 stock index returns is the surprisingly regular rise in the pre-crisis period (see Figs. 6a and 7). Leiss et al. [50] find that option-implied returns may be nicely described by a linear model as in Eq. (3) during the period from January 2003 to June 2007 (p-value < 0.001, $R^2 = 0.82$). The same does not hold true for the post-crisis period (p-value < 0.001, but with a coefficient of determination of $R^2 = 0.20$). The estimated linear coefficients are $\gamma = 0.01$ % and $\gamma = 0.003$ % per trading day before and after the crisis, respectively.

In order to reveal a similar trend in realized returns one needs to filter out short-term fluctuations, as they show a less regular behavior than returns expected under $\mathbb{Q}$. For this, we choose the exponentially weighted moving average (EWMA), an

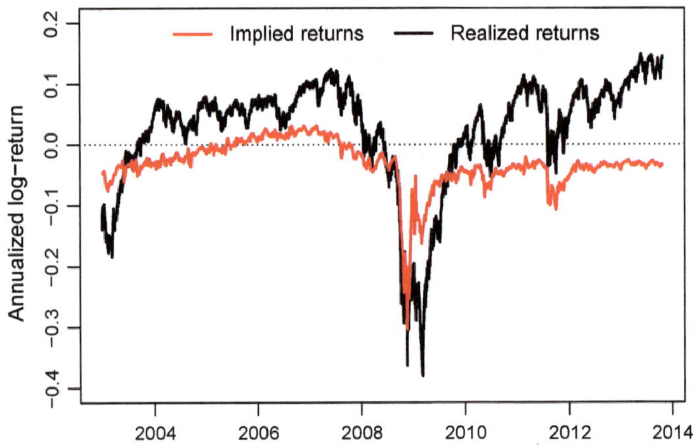

Fig. 7 Exponential weighted moving average of annualized realized and option-implied S&P 500 returns [50]

infinite impulse response filter with exponentially decreasing weights [36]. The results presented here are robust with respect to using other filters such as the local polynomial smoothing by Savitzky and Golay [66] as well as the choice of the smoothing factor. Figure 7 shows realized S&P 500 returns smoothed over 750 business days. The similarity between both time series in the pre-crisis period is remarkable and implies super-exponentiality both in expected and realized returns. As happens so often, the transient period of faster-than-exponential growth ends in a crash.

3.4 Granger-Causality Between Expected Returns and T-Bill Rates

Many monetary economists argue that the Global Financial Crisis of 2008 has led the Federal Reserve to take a new and more active part in steering financial markets [58]. Leiss et al. [50] employ Granger causality tests to analyze the mutual relationship between the market's return expectations and the Fed's monetary policy [31]. The two time-series of study are the first differences

$$SP_t = r_t^{\mathbb{Q}} - r_{t-1}^{\mathbb{Q}}, \quad TB_t = y_t - y_{t-1}. \tag{18}$$

where $r_t^{\mathbb{Q}}$ is the option-implied return (17) and y_t is the 3-month Treasury Bill yield at trading day t, respectively. Also, both time series are standardized to mean of zero and variance of one. Table 2 presents the results of a bidirectional Granger causality test with the null hypothesis of no Granger causality per subperiod identified in Sect. 3.2, respectively. Prior to the crisis, Leiss et al. [50] do not find signs of T-Bill yields Granger-causing expected stock returns at any lag, whereas for the other

Table 2 This table reports the results of a Granger-causality test of option-implied S&P 500 returns and Treasure Bill yields by sub-period [50]

Pre-crisis

	S&P Granger-causes T-Bill		T-Bill Granger-causes S&P	
Lag	F-ratio[a]	Degrees of freedom	F-ratio[a]	Degrees of freedom
5	2.72*	5, 926	0.23	5, 926
50	0.84	50, 791	0.82	50, 791
100	0.82	100, 641	0.92	100, 641
150	0.92	150, 491	0.77	150, 491
200	0.86	200, 341	0.87	200, 341
250	0.95	250, 191	0.83	250, 191

Post-crisis

	S&P Granger-causes T-Bill		T-Bill Granger-causes S&P	
Lag	F-ratio[a]	Degrees of freedom	F-ratio[a]	Degrees of freedom
5	1.95*	5, 942	0.56	5, 942
50	0.69	50, 807	1.37*	50, 807
100	0.79	100, 657	1.55**	100, 657
150	1.07	150, 507	1.32*	150, 507
200	1.16	200, 357	1.23*	200, 357
250	1.06	250, 207	1.18	250, 207

Note *$p < 0.1$; **$p < 0.01$; ***$p < 0.001$
[a]Refers to the F-test for joint significance of the lagged variables

direction the null of no Granger-causality from option-implied returns to T-Bill yields is rejected at a lag of $m = 5$ trading days. This means that the Federal Reserve played a rather passive, responding role between January 2003 and June 2007, which is line with previous works [32, 80].

The post-crisis analysis shows a similar mutual relationship at a short lag ($m = 5$), i.e. option-implied returns Granger-causing yields on Treasury Bills with the inverse not holding true. However, a significant change can be observed at longer time lags of $m = 50, 100, 150$ and 200 trading days. Here, Leiss et al. [50] report clear evidence of T-Bill yields Granger-causing option implied returns expected under risk-neutrality with the inverse not holding true. This confirms the generally held view that financial markets accepted the Federal Reserve to take an active, interventionist role paving the road to recovery.[6]

[6]As Leiss et al. [50] report, analyses of Granger causality including realized returns yield no comparable results.

4 Conclusion

In this chapter, we reviewed, discussed and extended two of our works on super-exponentiality in the context of finance. Super-exponentially growing asset prices are well documented empirically, and are of particular interest, as they are often associated with market instabilities, turbulence and bubbles. The studies presented here contribute to the arguably still incomplete understanding of faster-than-exponential growth by approaching it from two new perspectives. In the first part, we introduced the agent-based model of a financial market by Kaizoji et al. [46]. Building on a standard framework, the market was populated by two types of agents. The interplay and mutual interactions among fundamentalists on the one hand, and chartists on the other, created price patterns that resemble those of many financial markets. The dominance of herding due to social imitation over idiosyncratic investment decisions was found to be a key driver for super-exponential price paths. In particular, this transient explosive dynamic equaled the one earlier observed in controlled laboratory experiments. Moreover, we discussed an intriguing extension of the model by Philipp [62] introducing a third type of investors to the market, who specifically make use of the well established LPPLS framework for bubble detection. Instead of bubble mitigation due to arbitraging the transient unsustainable growth patterns, their presence is found to lead to bubbles with even higher prices.

In the second part, following Leiss et al. [50], we empirically estimated the distribution of return expectations of investors. Using quotes on financial options on the S&P 500 stock index, we constructed risk-neutral probability distributions of the representative investor for the decade 2003–2013. We presented an evaluation of the moments, tail characteristics and implied expected returns, that suggested two paradigm shifts due to the onset and end of the Global Financial Crisis of 2008. In particular, we found that investors exhibited super-exponential growth expectations during the pre-crisis period until June 2007. A Granger causality test suggested that the monetary policy of the Federal Reserve due to the market crash in 2008 had a profound and leading impact on the stock market, while this was not true prior to the crisis.

Both studies offer many opportunities for extensions and further research. The set-up by Philipp [62] is promising, but even outside of the simplified model world it is not clear how to optimally use the LPPLS methodology for investing. A refined and possibly heterogeneous set of LPPLS investment rules will lead to a more complex impact on the market price. This involves both enter and exit strategies for positions in the risky asset during a bubble and in phases of sustainable growth. As the agent-based model only exhibits approximate log-periodicity, this type of trader may also be equipped with a different methodology, that focuses only on the super-exponentiality of bubbles.

Moreover, one may use options data to analyze further transient price patterns, e.g. in other indices or individual stocks and generally for bubbles, that are smaller than the one which led to the Global Financial Crisis of 2008. This would allow to systematically evaluate the performance of various risk measures around the build-up of market instabilities. Besides the traditionally used risk measures such as value at risk, expected shortfall or volatility, Foster and Hart [27] introduce theoretical work on an intriguing new measure guaranteeing no-bankruptcy under the condition that the return probability distribution is known. Obviously, this does not hold true in practice and it is an open question, whether forward-looking return probability distributions constructed from options data represent an adequate information base for the measure by Foster and Hart [27].

Furthermore, as compared to equity markets, algorithms and quants have been taking over the more complex options exchanges only recently. Analyses on stock markets suggest, that machine trading has had a profound impact that goes beyond mere speed, but is accompanied by a new behavioral regime filled with subsecond extreme events [45]. This may be ever more true for structured financial products, as their complexity was found to introduce inherent instability to the market [3]. It will be very interesting to study the effects of the changing nature of markets due to algorithmic trading on price dynamics, market stability and systemic risks.

Acknowledgements I would like to thank Heinrich Nax and participants of the *Systemic Social and Financial Risk Workshop* of ETHZ/UZH for helpful comments as well as the ETH Risk Center and SNF grant *The Anatomy of Systemic Financial Risk* for financial support.

References

1. Akerlof, G.A., Shiller, R.J.: Animal Spirits: How Human Psychology Drives the Economy, and Why it Matters for Global Capitalism. Princeton University Press
2. Al-Anaswah, N., Wilfling, B.: Identification of speculative bubbles using state-space models with markov-switching. J. Bank. Finan. **35**(5), 1073–1086 (2011)
3. Battiston, S., Caldarelli, G., Georg, C.-P., May, R., Stiglitz, J.: Complex derivatives. Nat. Phys. **9**(3), 123–125 (2013)
4. Birru, J., Figlewski, S.: The Impact of the Federal Reserve's Interest Rate Target Announcement on Stock Prices: A Closer Look at How the Market Impounds New Information. New York University, Stern School of Business (2010)
5. Black, F.: Noise. J. Finan. **41**(3), 529–543 (1986)
6. Black, F., Scholes, M.: The pricing of options and corporate liabilities. J. Polit. Econ. 637–654 (1973)
7. Blanchard, O.: Speculative bubbles, crashes and rational expectations. Econ. Lett. **3**, 387–389 (1979)
8. Blanchard, O., Watson, M.: Bubbles, rational expectations and speculative markets. In: Wachtel, P. (ed.) Crisis in Economic and Financial Structure: Bubbles. Bursts and Shocks (1982)
9. Breitung, J., Kruse, R.: When bubbles burst: econometric tests based on structural breaks. Stat. Pap. **54**(4), 911–930 (2013)
10. Brock, W.A., Hommes, C.H.: A rational route to randomness. Econometrica 1059–1095 (1997)
11. Brock, W.A., Hommes, C.H.: Heterogeneous beliefs and routes to chaos in a simple asset pricing model. J. Econ. Dyn. Control **22**(8), 1235–1274 (1998)
12. Chiarella, C., Dieci, R., Gardini, L.: Asset price and wealth dynamics in a financial market with heterogeneous agents. J. Econ. Dyn. Control **30**(9), 1755–1786 (2006)
13. Chiarella, C., Dieci, R., He, X.-Z.: Heterogeneity, market mechanisms, and asset price dynamics. In: Schenk-Hoppé, K.R., Hens, T. (eds.) Handbook of Financial Markets: Dynamics and Evolution, Handbooks in Finance, pp. 277–344. North-Holland, San Diego (2009)

14. Chiarella, C., He, X.-Z.: Asset price and wealth dynamics under heterogeneous expectations. Quant. Finan. **1**(5), 509–526 (2001)
15. Chow, G.C.: Tests of equality between sets of coefficients in two linear regressions. Econometrica: J. Econom. Soc. 591–605 (1960)
16. De Long, J.B., Shleifer, A., Summers, L.H., Waldmann, R.J.: Noise trader risk in financial markets. J. Polit. Econ. 703–738 (1990a)
17. De Long, J.B., Shleifer, A., Summers, L.H., Waldmann, R.J.: Positive feedback investment strategies and destabilizing rational speculation. J. Finan. **45**(2), 379–395 (1990b)
18. Delbaen, F., Schachermayer, W.: A general version of the fundamental theorem of asset pricing. Mathematische Annalen **300**(1), 463–520 (1994)
19. Embrechts, P., Resnick, S.I., Samorodnitsky, G.: Extreme value theory as a risk management tool. North Am. Actuar. J. **3**(2), 30–41 (1999)
20. Engsted, T., Pedersen, T.Q.: The dividend-price ratio does predict dividend growth: international evidence. J. Empir. Finan. **17**(4), 585–605 (2010)
21. Fama, E.F.: Efficient capital markets: a review of theory and empirical work*. J. Finan. **25**(2), 383–417 (1970)
22. Federal Reserve Bank of St. Louis: The financial crisis: a timeline of events and policy actions. http://timeline.stlouisfed.org/pdf/CrisisTimeline.pdf (2009)
23. Figlewski, S.: Estimating the implied risk neutral density. In: Bollerslev, T., Russell, J., Watson, M. (eds.) Volatility and Time Series Econometrics. Oxford University Press, Oxford (2010)
24. Figlewski, S.: What is risk neutral volatility? (2012). Available at SSRN 2183969
25. Filimonov, V., Sornette, D.: Quantifying reflexivity in financial markets: toward a prediction of flash crashes. Phys. Rev. E **85**(5), 056108 (2012)
26. Filimonov, V., Sornette, D.: A stable and robust calibration scheme of the log-periodic power law model. Phys. A: Stat. Mech. Appl. **392**(17), 3698–3707 (2013)
27. Foster, D.P., Hart, S.: An operational measure of riskiness. J. Polit. Econ. **117**(5), 785–814 (2009)
28. Fouque, J., Papanicolaou, G., Sircar, K.: Derivatives in Financial Markets with Stochastic Volatility. Cambridge University Press (2000)
29. Friedman, M.: The Case for Flexible Exchange Rates. University of Chicago Press (1953)
30. Gibrat, R.: Les inégalités économiques. Recueil Sirey (1931)
31. Granger, C.W.J.: Investigating causal relations by econometric models and cross-spectral methods. Econometrica **37**(3), 424–438 (1969)
32. Guo, K., Zhou, W.-X., Cheng, S.-W., Sornette, D.: The US stock market leads the federal funds rate and treasury bond yields. PLoS ONE **6**(8), e22794 (2011)
33. Hamilton, J.D.: A new approach to the economic analysis of nonstationary time series and the business cycle. Econometrica 357–384 (1989)
34. Hamilton, J.D., Raj, B.: New directions in business cycle research and financial analysis. Empir. Econ. **27**(2), 149–162 (2002)
35. Harras, G., Tessone, C.J., Sornette, D.: Noise-induced volatility of collective dynamics. Phys. Rev. E **85**(1), 011150 (2012)
36. Holt, C.C.: Forecasting seasonals and trends by exponentially weighted moving averages. Int. J. Forecast. **20**(1), 5–10 (2004)
37. Homm, U., Breitung, J.: Testing for speculative bubbles in stock markets: a comparison of alternative methods. J. Finan. Econom. **10**(1), 198–231 (2012)
38. Hong, H., Kubik, J.D., Stein, J.C.: Thy neighbor's portfolio: word-of-mouth effects in the holdings and trades of money managers. J. Finan. **60**(6), 2801–2824 (2005)
39. Hüsler, A., Sornette, D., Hommes, C.H.: Super-exponential bubbles in lab experiments: evidence for anchoring over-optimistic expectations on price. J. Econ. Behav. Organ. **92**, 304–316 (2013)
40. Ising, E.: Beitrag zur Theorie des Ferromagnetismus. Zeitschrift für Physik A Hadrons and Nuclei **31**(1), 253–258 (1925)
41. Jackwerth, J.C.: Option-Implied Risk-Neutral Distributions and Risk Aversion. Research Foundation of AIMR Charlotteville (2004)

42. Jiang, Z.-Q., Zhou, W.-X., Sornette, D., Woodard, R., Bastiaensen, K., Cauwels, P.: Bubble diagnosis and prediction of the 2005–2007 and 2008–2009 chinese stock market bubbles. J. Econ. Behav. Organ. **74**(3), 149–162 (2010)
43. Johansen, A., Ledoit, O., Sornette, D.: Crashes as critical points. Int. J. Theoret. Appl. Finan. **3**(2), 219–255 (2000)
44. Johansen, A., Sornette, D., Ledoit, O.: Predicting financial crashes using discrete scale invariance. J. Risk **1**(4), 5–32 (1999)
45. Johnson, N., Zhao, G., Hunsader, E., Qi, H., Johnson, N., Meng, J., Tivnan, B.: Abrupt rise of new machine ecology beyond human response time. Sci. Rep. **3** (2013)
46. Kaizoji, T., Leiss, M., Saichev, A., Sornette, D.: Super-exponential endogenous bubbles in an equilibrium model of fundamentalist and chartist traders. J. Econ. Behav. Organ. (2015)
47. Kirman, A.: Ants, rationality, and recruitment. Q. J. Econ. 137–156 (1993)
48. Kyle, A.S.: Continuous auctions and insider trading. Econometrica: J. Econom. Soc. 1315–1335 (1985)
49. Lammerding, M., Stephan, P., Trede, M., Wilfling, B.: Speculative bubbles in recent oil price dynamics: evidence from a bayesian markov-switching state-space approach. Energy Econ. **36**, 491–502 (2013)
50. Leiss, M., Nax, H.H., Sornette, D.: Super-exponential growth expectations and the global financial crisis. J. Econ. Dyn. Control (forthcoming) (2015)
51. Lin, L., Sornette, D.: Diagnostics of rational expectation financial bubbles with stochastic mean-reverting termination times. Eur. J. Fin. (2011). doi:10.1080/1351847X.2011.607004
52. Lorenz, J., Rauhut, H., Schweitzer, F., Helbing, D.: How social influence can undermine the wisdom of crowd effect. Proc. Natl. Acad. Sci. **108**(22), 9020–9025 (2011)
53. Lux, T.: Herd behaviour, bubbles and crashes. Econ. J. 881–896 (1995)
54. Lux, T.: Stochastic behavioral asset-pricing models and the stylized facts. In: Schenk-Hoppé, K.R., Hens, T. (eds.) Handbook of Financial Markets: Dynamics and Evolution, Handbooks in Finance, pp. 161–215. North-Holland, San Diego (2009)
55. Lux, T., Marchesi, M.: Scaling and criticality in a stochastic multi-agent model of a financial market. Nature **397**(6719), 498–500 (1999)
56. Mandelbrot, B.: The variation of certain speculative prices. J. Bus. **36**(4), 394–419 (1963)
57. Mantegna, R.N., Stanley, H.E.: Scaling behaviour in the dynamics of an economic index. Nature **376**(6535), 46–49 (1995)
58. Mehrling, P.: The New Lombard Street: How the Fed Became the Dealer of Last Resort. Princeton University Press (2010)
59. Orléan, A.: Efficience, finance comportementale et convention: une synthèse théorique. Les crises financieres, Rapport de Conseil d?Analyse Economique **50**, 241–270 (2004)
60. Page, E.: Continuous inspection schemes. Biometrika **41**(1/2), 100–115 (1954)
61. Pareto, V.: Cours d'économie politique. Macmillan, Paris (1897)
62. Philipp, D.: Influence of LPPL traders on financial markets. Master's thesis, Eidgenössische Technische Universität Zürich, Switzerland (2015)
63. Phillips, P.C.B., Shi, S.-P., Yu, J.: Testing for multiple bubbles 1: historical episodes of exuberance and collapse in the S&P 500. Cowles Foundation Discussion Paper No. 1843 (2012)
64. Phillips, P.C.B., Wu, Y., Yu, J.: Explosive behavior in the 1990s Nasdaq: when did exuberance escalate asset values? Int. Econ. Rev. **52**(1), 201–226 (2011)
65. Saichev, A., Malevergne, Y., Sornette, D.: Theory of Zipf's Law and Beyond (2009)
66. Savitzky, A., Golay, M.J.: Smoothing and differentiation of data by simplified least squares procedures. Anal. Chem. **36**(8), 1627–1639 (1964)
67. Scott, A., Knott, M.: A cluster analysis method for grouping means in the analysis of variance. Biometrics **30**(3), 507–512 (1974)
68. Shilling, A.G.: Scoreboard: A frank self-appraisal of where i went wrong and where I went right in the year just past. Forbes **151** (1993)
69. Sornette, D.: Why Stock Markets Crash: Critical Events in Complex Financial Systems. Princeton University Press (2003)

70. Sornette, D., Cauwels, P.: 1980–2008: The illusion of the perpetual money machine and what it bodes for the future. Risks **2**(2), 103–131 (2014a)
71. Sornette, D., Cauwels, P.:. Financial bubbles: mechanisms and diagnostics. Swiss Finance Institute Research Paper (14-28) (2014b)
72. Sornette, D., Johansen, A.: A hierarchical model of financial crashes. Phys. A: Stat. Mech. Appl. **261**(3), 581–598 (1998)
73. Sornette, D., Woodard, R.: Financial Bubbles, Real Estate Bubbles, Derivative Bubbles, and the Financial and Economic Crisis. Springer (2010)
74. Sornette, D., Woodard, R., Yan, W., Zhou, W.-X.: Clarifications to questions and criticisms on the Johansen-Ledoit-Sornette bubble model. Phys. A: Stat. Mech. Appl. **392**(19), 4417–4428 (2013)
75. Sornette, D., Woodard, R., Zhou, W.-X.: The 2006–2008 oil bubble: evidence of speculation, and prediction. Phys. A: Stat. Mech. Appl. **388**(8), 1571–1576 (2009)
76. Soros, G.: The New Paradigm for Financial Markets: The Credit Crisis of 2008 and What it Means. Basic Books (2008)
77. Wyart, M., Bouchaud, J.-P.: Self-referential behaviour, overreaction and conventions in financial markets. J. Econ. Behav. Organ. **63**(1), 1–24 (2007)
78. Yan, W., Rebib, R., Woodard, R., Sornette, D.: Detection of crashes and rebounds in major equity markets. Int. J. Portfolio Anal. Manage. **1**(1), 59–79 (2012)
79. Zhou, W.-X., Sornette, D.: 2000–2003 real estate bubble in the UK but not in the USA. Phys. A: Stat. Mech. Appl. **329**(1), 249–263 (2003)
80. Zhou, W.-X., Sornette, D.: Causal slaving of the US treasury bond yield antibubble by the stock market antibubble of August 2000. Phys. A: Stat. Mech. Appl. **337**(3), 586–608 (2004)
81. Zhou, W.-X., Sornette, D.: Is there a real-estate bubble in the us? Phys. A: Stat. Mech. Appl. **361**(1), 297–308 (2006)

Part II
The View from Inside

Methods and Finance. A View From Inside

Emiliano Ippoliti

Abstract The view from inside maintains that not only to study and understand, but also to profit from financial markets, it is necessary to get as much knowledge as possible about their internal 'structure' and machinery. This view maintains that in order to solve the problems posed by finance, or at least a large part of them, we need first of all a qualitative analysis. Rules, laws, institutions, regulators, the behavior and the psychology of traders and investors are the key elements to the understanding of finance, and stock markets in particular. Accordingly, data and their mathematical analysis are not the crucial elements, since data are the output of a certain underlying structure of markets and their actors. The underlying structure is the ultimate object of the inquiry. This chapter examines how the view from inside raises, and deals with, critical issues such as markets failure, information disclosure, and regulation (Sect. 2), the notion of data (Sect. 3), performativity (Sect. 4), and the study of micro-structures (Sect. 5).

1 The Inside View: An Overview

The view from inside and the view from outside make different assumptions about critical ontological and methodological issues on finance, in particular the role of mathematical modeling, the methods to be employed, and the nature and set of variables to employ in order to achieve a better understanding of the financial systems.

The inside view argues that not only to study and understand, but also to profit from markets, it is necessary to get as much knowledge as possible about their internal 'structure' and machinery. This view maintains that in order to solve the problems posed by finance, or at least a large part of them, we need first of all a qualitative analysis. Rules, laws, institutions, regulators, the behavior and the psychology of traders and investors are the key elements to the understanding of finance, and stock markets in particular. In its strong version, this view maintains that finance is a black box—plenty of dark zones and opacities. Accordingly, data

E. Ippoliti (✉)
Department of Philosophy, Sapienza University of Rome, Rome, Italy
e-mail: emi.ippolit@gmail.com

E. Ippoliti and P. Chen (eds.), *Methods and Finance*,
Studies in Applied Philosophy, Epistemology and Rational Ethics 34,
DOI 10.1007/978-3-319-49872-0_7

and their mathematical analysis are not the crucial element, since data are the output of a certain underlying structure of markets and their actors. The underlying structure is the ultimate object of the inquiry. This view conceives the role of the data in a very restricted fashion: since the data express *historical* processes and structures, they have to be continually and carefully contextualized: we can treat them as homogenous only under very specific conditions. This fact implies that a large part of the mathematical-statistical analysis of markets would be inappropriate, or very partial. In turn this prevents us from using a given dataset as a tool to understand, and possibly make forecasts, about other datasets that are outcomes of different underlying structures, both at a macro and at micro timescale. A lot of mathematical tools employed in finance need to assume continuity and homogeneity between datasets and this fact is not justified from an internalist viewpoint. A hypothesis generated staring from a given dataset and structure, for instance the Efficient Market Hypothesis (EMH), is largely ineffective for the understanding of markets' dynamics of other dataset, i.e. today's dynamics, since the EMH is an attempt of explaining a specific dataset, derived from 1960s markets, with a particular structure that is no longer at work. Just to mention few aspects, the data for prices and volumes were not collected in real time in the '60s, were all low frequency in nature, and the markets were regulated environments with controlled and limited cross market transactions. Of course we can still see 'similar' patterns and graphs in the 'data', but since they are the outputs of internal machineries very different, their comparison would be useless or misleading most of the times.

As a matter of fact, a better understanding of the financial systems can be achieved by an examination of their *inner* components, starting from the features of their agents or actors—in particular their behavior and psychology. Behavioral finance (BF) is a stock example in this sense. BF (see [3]) stems from the recent advancements in cognitive psychology and evolutionary psychology [5, 6, 9, 11], endowed with the notions of ecological rationality and massive modularity of our brain [4, 10, 12]. These two branches of psychology shape the main theories put forward in BF, respectively the Heuristics & Biases approach (H&B, see [2, 16, 17, 24, 25]) and the Fast and Frugal approach (F&F, see [13–15]). They provide a different characterization of decision-making and rationality, even if they share the basic idea that emotions and psychological features affect the decision-making of traders and investors. The H&B tradition argues that emotions and psychological features (e.g. loss aversion) generates heuristics and biases leading to mistakes and errors, that is a violation of optimization process. The H&B theorists basically examine how, and to what extent, the use of heuristics interferes with the selection of options that would maximize the classics expected value of conventional rational choice theory. In effect, the heuristics produces a miscomputation of probabilities that a choice will have certain consequences or the misevaluation of the possible end states. Of course, the H&B tradition recognizes the problematic role of heuristics: the employment of a heuristics does not necessary lead to errors and, as a matter of fact, it produces good inferences or decisions in most of the cases. What leads to mistakes is that "the correct conclusion doesn't always follow from the most readily processed cues and reactions. The additional, harder-to-process factors and reactions—those that can generally be ignored with little cost to accurate decision making—may sometimes turn out to be crucial" [18, p. 6].

On the other side, the F&F tradition stresses how a set of heuristic tools can even be better than optimization in specific cases. In effect, the F&F theorists examine heuristics in the sense of a problem-solving techniques that allows us to make decisions that achieve the ends of an organism in a given 'ecological niche', regardless of the fact that techniques is 'rational' in a strict sense. So the achievements, and not 'biases' or 'errors', are their main focus.

More in detail, the H&B approach maintains that heuristics and biases work in such a way to produce decisions and inferences, and accordingly prices of financial securities, that contradict optimal rationality—as formalized by the Expected Value Theory (EVT). Of course the consequences of the use of these heuristics and biases are reflected in prices patterns: they can trigger particular phenomena, like crashes and bubbles, which are seen as anomalies or failures by other approaches. In this view "the most significant contribution of the H&B school has been to detail a novel sort of market failure" [18, p. 6]. In effect, the H&B focuses *not* on the *external* limitations causing bad choices, such as the lack of information about the available options, but on the *internal* limitations of actors, which prevent them to use in a proper way even the information that they already have.

On the other side, the F&F theorists turn out to support the rational choice approach' idea that people have reliable mechanisms to make 'good' choices, even though the mechanisms they provide are different from the ones identified by the rational choice theory. These mechanisms are adaptive in nature, modular (even massively), ecologically rational, and for this reason they would provide tools to make decisions and drawing inferences that can even be better than the options provided by the EVT. In effect the F&F tradition emphasizes how decisions and inferences are often made considering a single dominant factor, that is in a lexical, non-compensatory fashion.[1]

Accordingly the adoption of one of the two approaches is rich of consequences for a inside viewpoint on finance, for they do disagree not only about whether heuristics can "better be seen as a frequent source of error or as the basic building block of intelligence and functional decision making" [18, p. 7] but they also disagree also about "why heuristics lead to mistakes when they do lead to mistakes, think differently about what it means to behave rationally, [...] how the use of particular heuristics emerges, [...] the processes by which we reach judgments when reasoning heuristically, and [...] whether people are less prone to use heuristics when they have certain background traits (intelligence, disposition) or have more time or cognitive resources to reach a decision" (*Ibid.*). In particular, H&B and F&F have different implications in terms of policy: they suggest different strategies to policymakers.

[1]A decision is 'lexical' when somebody prefers *A* instead of *B* *because* *A* is considered better along a single dominant feature, without considering other 'compensating' properties that *B* might have in comparison to *A*. For example we chose a pizza *A* over *B* in a lexical fashion when *A* is selected just because it is cheaper, without looking at other features.

2 Markets Failure, Information Disclosure, and Regulation

The adoption of one of these two approaches leads to two very different views on critical features of markets, such as failure, information disclosure, and regulation. First of all, H&B has displayed new characteristics of markets and regulation, namely forms of failures.

In the former case, that is failure of markets, the H&B approach has identified frameworks in which an agent endowed with enough information about a domain, the *external* world, do not *internally* process those pieces of information in a proper way, that is in a fashion that allows the agents to draw the correct inference about either the *external* world or their preferences. The agent *fail* in the elaboration of information for internal reasons, and consequently the market fail for *internal* reason. The agent, and not the information, is the root and reason of the failure.

In the latter case, the H&B approach has highlighted new kinds of regulatory failure. As concern financial markets, this approach shows that 'anomalies' in these markets (a stock example is a 'bubble'–or a crash) emerge because agents are subject to a class of mistakes and biases, coming for example from sensitivity to the manner of elicitation, and hence a greater regulatory intervention should be implemented in those markets. The bottom line: the market decisions should be often entrusted to experts in the field.

On the other side, the F&F tradition has provided several findings and arguments that show that when information is disclosed to agents in a fuller way, they will face an information overload leading to mistakes, rather than better decisions. A clarification is needed here. The F&F theory does *not* argue that unregulated markets, and the use of F&F heuristics, will permit agents to assess their options in an ideal way.[2] As a matter of fact, the F&F school points out that a lot of massive regulation aiming at protecting agents from making irrational decisions, turns out to interfere with their 'natural' skill to gather and use decision-improving information. In particular, typical disclosure policies are misleading: whilst improving the *form* in which information is presented is effective most of the times, increasing the amount of information that agents receive is ineffective to a large extent since too many pieces of information makes it harder the identification of a single, best lexical cue.

As a consequence, on one side, the H&B tradition turns out to support a 'soft paternalistic' approach, in the sense that if we commit to it, then agents (traders, consumers, voters) need a help in avoiding mistakes—and a way to do this is to cede decision-making authority to 'experts' who are less subject to these mistakes. Thus this approach believes that it is right to interfere with agents' immediate

[2]Let us consider, for instance, one of the most basic fast and frugal heuristics—the recognition heuristic. In this case, the heuristics could be exploited in several ways. As a stock example, let us consider an advertiser that work to ensure that her product or brand be merely recognized, exploiting the fact that consumers often jump to the conclusion that recognition and product desirability are connected.

freedom, even though it maintains that they should have their autonomy, and the right to have atypical preferences, and to manifest them in a way that mandates would not. On the other side the F&F approach supports a different view: agents employ strategies that work well for them, even though these strategies contradict the rational choice theory, and therefore they should have more freedom—and they do not need mandates most of the times.

3 Data: An Inside View

The view from inside provides also new perspectives on the role and use of data. In effect, an interesting finding of the F&F school is the existence of *additional* problems with data, that is *internal* difficulties. These problems, more precisely, arise from the quantity and the form of data that conflict with the 'natural' way of agents of processing them. This is also one of the main points of conflict between H&B and F&F approaches.

As concerns the quantity of data, the F&F school argues that too much information sent to agents will not improve their ability to make good decisions, since some of these *mandated* pieces of information will be hard (expensive, in terms of time and cognitive skills) to process. Instead of producing a better decision or inference, this information will act as distractors or barriers, producing a poorer decision: they will interfere with an agent's attempt to locate the best piece of information (while applying heuristics like "take the best" or "follow the crowd"). So a reduction of data, or better the information put into them, will ease the inferential or the decision-taking process. In the specific case of markets, the F&F school argues that they originate the useful cues most of the times, but a heavy or improper regulation can obscure them and push agents to ignore all the information. The bottom line: an agent will act as fully ignorant. In some case, the situation gets even worse, for agents exposed to too many mandated disclosures could try to combine multiple cues, making compensatory decision with counterproductive effects.[3]

As concerns the form of data, the stock example provided by the F&F school is the use of the frequentist form instead of the probabilistic one in the assessment of risk (see [13]). In effect the F&F approach has experimentally shown that the use of probabilities and percentages generates an overestimation of risks, since agents tend to neglect the base-rate information. Since the elaboration of information is sensitive to their form of presentation to an agent, a more appropriate form the information put into them will ease the inferential or the decision-taking process. It is important to emphasize that the inferences or the decisions made by those exposed to frequencies are different but not necessarily superior to those made by

[3]Common examples are overfitting data or integrating incommensurable values.

those exposed to probabilities or percentages. In some case, the adoption of a more 'natural' presentation in the data does not produce effect at all.

4 Performativity

The inside view also arises a performativity problem [7, 8, 19, 20], like the outside view. There is a way in which the *internal* financial models and theories influence and shape the systems they seek to understand. More precisely, the BF and its two main branches involve a performative issue, or better a counter-performative one. Tellingly, the adoption of this conceptual framework changes the behavior of those who practice it, and accordingly the dynamics of the markets, by making its practitioners resemble less its account and description. In other words, the use of BF will produce agents, and market's dynamics, which are less and less emotional, attentive, calibrated. Thus, as noted by Alex Preda in his paper for this volume, we face a seemingly paradoxical situation: BF would make real market agents closer to the opposite approach, which is shaped by the ideal of emotionless, fully attentive, perfectly calibrated market agents.

5 Micro-structures

A radical change of scale and of focus are at the core of another recent *inside* approach: the one based on *microstructures* (see e.g. [21–23, 26]). While BF is focused on the features of *human* decision-making, markets microstructures theory basically examines at a finer grain the *order flow*, that is how the transactions are executed, and data and information are constructed, queued, and disclosed.[4] If traders, mostly humans, are still the basic unit of inquiry in the BF approach, in markets microstructures approach things are very different. First of all traders are silicon, not human. Second, market data are not the trades, but their internal machinery, that is the underlying orders. This change is mainly due to the expansion of algorithmic trading and, in particular, one of its bran ramifications, namely HFT.

The increasing amount of findings about how transaction costs, prices, quotes, volume, and trading behavior in general are affected by the working processes of a market is changing the way of understanding and acting on financial systems and their dynamics. The emergence of events like flash crashes, that is the plummet and

[4]O'Hara [22] defines it as the "study of the process and outcomes of exchanging assets under explicit trading rules. While much of economics abstracts from the mechanics of trading, microstructure literature analyzes how specific trading mechanisms affect the price formation process.

the rebound of a asset in few minutes, or mini flash crash,[5] are connected to the changes in the micro structures of the markets.

The understanding of these events requires a deconstruction of data about trading rules and time (see e.g. [1]) that can be effectively approached only by means of a micro level approach.

A striking feature of the study of microstructures is that this process is not necessarily human, since computers can handle an entire order flow: so a financial agent, in this case, can be entirely 'mechanical', an algorithm programmed to trade (place orders) or to recognize specific dynamics in the prices.

The study of microstructures (MS), first of all, is based on a change of the time scale. The focus here is on the micro scale, the very short-term price variations: the tick-by-tick variations that take place even in terms of milliseconds. In this sense the dataset employed are not the usual financial data.

Secondly, it is based on a change of focus, that is on process and not on variables. It examines the several *rules of execution* of the buy or sell orders on a particular exchange venue. The basic idea behind MS approach is that the 'game' of financial markets is played by agents who will employ strategies also at micro level: they exploit the features of trading rules of a particular exchange venue, in combination with pieces of information about other orders and their queues and timestamp, in order to shape a strategy that will offer advantages over their opponents. Moreover they will seek these opportunities without revealing their intentions or all the information at their disposal. The basic tenet is that the final outcome of the order flow, that is prices and their dynamics, not only can understood in terms of the strategies of the financial agents involved in the game, but that they offer a way to interpret these strategies and, eventually, reveal them.

Moreover, MS approach is sensitive to technology, since the rules of execution of orders and the strategies depend on the platform employed to place them. For instance, in case of an electronic platform, the primary concern will be sequence of price and volume data, and their time.

References

1. Aldrich, E.M., Grundfest, J., Laughlin, G.: The Flash Crash: A New Deconstruction, SSRN-id2721922 (2016)
2. Barberis, N., Huang, M., Santos, T.: Prospect theory and asset prices. Q. J. Econ. **116**(1), 1–53 (2001)

[5]Mini Flash Crashes were first identified by Nanex Llc., and defined as follows.

A down crash has to be such that the stock price change:

(i) it has to tick down at least 10 times before ticking up,

(ii) price changes have to occur within 1.5 s,

(iii) price change has to exceed –0.8 %.

An up crash has to be such that the stock price change:

(i) it has to tick up at least 10 times before ticking down,

(ii) price changes have to occur within 1.5 s,

(iii) price change has to exceed 0.8 %.

3. Barberis, N., Thaler, R.: A survey of behavioral finance. In: Constantinides, G., Harris, M., Stulz, M. (eds.) Handbook of the Economics of Finance: Volume 1B, Financial Markets and Asset Pricing, pp. 1053–1128. Elsevier, North Holland (2003)
4. Barkow, J., Cosmides, L., Tooby, J. (eds.): The Adapted Mind: Evolutionary Psychology and the Generation of Culture. Oxford University Press, New York (1992)
5. Buss, D.: Evolutionary Psychology: The New Science of the Mind. Pearson/A and B, Boston (2004)
6. Buss, D. (ed.): The Handbook of Evolutionary Psychology. Wiley, Hoboken, NJ (2005)
7. Callon, M.: Introduction: the embeddedness of economic markets in economics. In: Callon, M. (ed.) The Laws of the Markets. Blackwell, Oxford (1998)
8. Callon, M.: What does it mean to say that economics is performative? In: MacKenzie, D., Muniesa, S. (eds.) Do Economists Make Markets? On the Performativity of Economics, pp. 310–357. Princeton University Press, Princeton (2007)
9. Confer, J.C., Easton, J.A., Fleischman, D.S., Goetz, C.D., Lewis, D.M.G., Perilloux, C., Buss, D.: Evolutionary psychology: controversies, questions, prospects, and limitations. Am. Psychol. **65**(2), 110–126 (2010)
10. Cosmides, L., Tooby, J.: Origins of domain specificity: the evolution of functional organization. In: Hirschfeld, L.A., Gelmen, S.A. (eds.) (2000) Mapping the Mind: Domain Specificity in Cognition and Culture, pp. 523–543. Cambridge University Press, Cambridge (1994)
11. Evan, D.: Introducing Evolutionary Psychology. Totem Books USA, Lanham, MD (2000)
12. Fodor, J.A.: Modularity of Mind: An Essay on Faculty Psychology. MIT Press, Cambridge, MA (1983)
13. Gigerenzer, G., Todd, P.M.: Simple Heuristics That Make Us Smart. Oxford University Press, New York (1999)
14. Gigerenzer, G., Hertwig, R., Pachur, T.: Heuristics: The Foundation of Adaptive Behavior. Oxford University Press, New York (2011)
15. Gigerenzer, G., Selten, R. (eds.): Bounded Rationality: The Adaptive Toolbox. Dahlem Workshop Reports. The MIT Press, Cambridge, MA (2001)
16. Kahneman, D.: Thinking, Fast and Slow. Macmillan, London (2011)
17. Kahneman, D., Tversky, A.: Prospect theory: an analysis of decision under risk. Econometrica **47**(2), 263–291 (1979)
18. Kelman, M.: The Heuristic Debate. Oxford University Press, Oxford (2011)
19. MacKenzie, D.: An engine, not a camera. How Financial Models Shape Markets. MIT Press, Cambridge (MA) (2006)
20. MacKenzie, D., Muniesa, F., Siu, L. (eds.): Do Economists Make Markets? On the Performativity of Economics. Princeton University Press, Princeton (2007)
21. Madhavan, A.: Market microstructure: a survey. J. Finan. Mark. **3**, 205–258 (2000)
22. O'Hara, M.: Market Microstructure Theory. Blackwell, Malden MA (1995)
23. O'Hara, M.: High Frequency Market Microstructure. J. Finan. Econ. **116**(2), 257–270 (2015)
24. Tversky, A., Kahneman, D.: Rational choice and the framing of decisions. J. Bus. Part 2: Behav. Found. Econ. Theory **59**(4), S251–S278 (1986)
25. Tversky, A., Kahneman, D.: Advances in prospect theory: cumulative representation of uncertainty. J. Risk Uncertain. **5**, 297–323 (1992)
26. Wood, R.A.: Market microstructure research databases. History and projections. J. Bus. Econ. Stat. **18**(2), 140–145 (2000)

Contemporary Finance as a Critical Cognitive Niche: An Epistemological Outlook on the Uncertain Effects of Contrasting Uncertainty

Tommaso Bertolotti and Lorenzo Magnani

Abstract Cognitive niche construction theory provides a new comprehensive account for the development of human cultural and social organization with respect to the management of their environment. Cognitive niche construction can be seen as a way of lessening complexity and unpredictability of a given environment. In this paper, we are going to analyze economic systems as highly technological cognitive niches, and individuate a link between cognitive niche construction, unpredictability and a particular kind of economic crises.

Keywords Cognitive niche construction · Technology · Financial crisis · Unpredictability

1 Introduction: Cognition Versus Unpredictability and Uncertainty

Complexity and Life, at every stage of evolution, are two notions that are never that far one from the other. The very origin of life on Earth, and the fact that so far there is no strong *evidence* of life having originated anywhere else in the Universe, are considered as the epitome of a random and unpredictable (and hardly reproducible) series of conditions.

This article was originally published as T. Bertolotti and L. Magnani (2015), Contemporary Finance as a Critical Cognitive Niche, *Mind & Society*, 14(2) 273–293. Special issue on "Complexity Modeling in Social Science and Economics", edited by Itzhak Aharon, Sacha Bourgeois-Gironde, and Yakir Levin.

T. Bertolotti (✉) · L. Magnani
Department of Humanities – Philosophy Section & Computational Philosophy Laboratory, University of Pavia, Piazza Botta, 6, 27100 Pavia, Italy
e-mail: bertolotti@unipv.it

L. Magnani
e-mail: lmagnani@unipv.it

E. Ippoliti and P. Chen (eds.), *Methods and Finance*,
Studies in Applied Philosophy, Epistemology and Rational Ethics 34,
DOI 10.1007/978-3-319-49872-0_8

The past century, also because of the development of quantum physics, has witnessed a growing awareness about how complexity and unpredictability are dominant constraints in ontogenesis and phylogenesis, that is affecting not only the development of species as we know them, but also of every single individual [19]. In other words, given an initial set of genetic codes, there is no way to predict their giving rise to a potentially infinite number of new species; similarly, given the genetic code of an embryo, there is no way to reliably predict (or compute) what the phenotypic expression will be in *n* years, because of the unpredictable effects of a highly unpredictable environment.

Indeed, from the perspective of system studies, life emerged and prospered in a non-deterministic, open system. Such systems are characterized by a significant level of uncertainty, in the sense that given current conditions future outcomes can be foreseen rightly to certain extents, or wrongly guessed, but they cannot be deterministically computed. This is essential to the nature of the system.

The relationship between an organism and its environment is indeed one of coping with the uncertainty of its system, and that—we contend—is what sparked the origin of the multifaceted phenomenon known as cognition. Any definition of cognition is conceptually human-centered, and has been only subsequently extended—at least possibly—to animals. This is why we suggest that adopting a barely-essential definition of cognition as the one offered by the *Stanford Philosophy Encyclopedia* might serve our scope, and let us rely on something that is not overly biased towards animals, but neither excessively human centered to begin with.

> Cognition is constituted by the processes used to generate adaptive or flexible behavior.

The adaptive behavior implied by cognitive capabilities is a response to the uncertainty of the environment in which the organism must survive.

We can wonder: what is, pragmatically speaking, the ultimate goal of cognition? If by introducing a difference between unpredictability and uncertainty we mean to stress the qualitative gap between, respectively, quantifiable (unpredictable) and unquantifiable (uncertain) risk, a stimulating way to answer might be the following: *cognition aims at predicting what is predictable, making predictable what is unpredictable, and making just unpredictable what is uncertain.* The two first thirds of this sentence are clear enough, as they frame the cognitive drive towards *making sense of one's surroundings* [22, Chap. 3]. The latter, conversely, considers how survival, at any level (from the humblest ant to the Wolf of Wall Street) is mostly about achieving second bests. In this sense, we can understand cognition as the strife to pull within the range of predictability what is uncertain.

With this respect, the form of inference best characterizing the organism's approach to its surroundings is *abduction*, which in this case accounts for the attempt (sometimes successful) to elaborate and enact hypotheses about the behavior of relevant objects in an organism's environment [23]. We shall limit our analysis of abduction to what is requested by our current goal, but it is interesting from the beginning to contrast it with a better known form of inference, that is *deduction*. This very brief example should make the distinction clear, and show why the notion is so relevant for our task.

- Consider a train running on a rail track. If you know the speed of the train, the path of the tracks and so on, it is relatively straightforward *ceteris paribus*[1] to calculate the position of the train at any given time. In a way, the position of the train can be *deduced* from its speed, its table and the geography of the track. If these are true, that is in accordance with the related state of things in the world, then the train will be where we compute it to be at a given time.
- Consider a lion chasing a gazelle. The lion must anticipate the prey's movement in order to tackle it to the ground and kill it. Still there is no way for the lion, or for any hunter, to *deduce* the position of the escaping prey at any given time. The lion must *perform and enact immediately* a quick appraisal on whether the gazelle will jerk right or left, basing on speed, maybe past experiences, terrain conformation, presence of other lions and so on: the future position of the gazelle is quite unpredictable, therefore an *abductive hypothesis*, enacted at once, is the best the lion can rely on to manage its pursuit.

What is the difference between the two systems described in the examples above? Standard models, ideologically assuming learning processes to be deterministic, would suggest that there is hardly any qualitative difference, and a complex model would explain the lion and gazelle system as well, making it predictable once the various factors are appropriately quantified. Translated into popular culture, this kind of assumption is precisely what lead to the bloodshed on Isla Nublar, in the novel and blockbuster Jurassic Park: such remark shows how common sense is quite clear about the difference between closed systems (more or less complicatedly deterministic), and open systems (uncertain and not necessarily quantifiable).[2] Going back to our examples, train circulation is a globally deterministic system inasmuch as it is entirely man-made: ruling out major break-ins of *life* in the system, for instance in the form of a suicidal will in the conductor, a human error, a terrorist attack etc., it is generally simple to foresee a schedule for the train also quantifying minor unpredictable disturbances. This is why, on the overall, the train schedules are quite reliable and indeed we usually experience pragmatically positive outcomes if we *assume* them to be reliable, and *behave* consequently.

Conversely, the lion chasing the gazelle requires an entirely different paradigm to be described. Indeed, even something as simple as chasing a gazelle is indeed an open system, characterized by a great degree of uncertainty. Not only elements of unpredictability are hard to quantify, but the uncertainty also concerns the individuation of relevant factors and their interactions. Physical systems, no matter how

[1]That is, assuming there is no accident, no high jacking, no passenger activates the emergency break, no excessive and unjustified delay, etc.

[2]Jurassic Park can almost be considered as a biogenetic thought experiment, given the amount of scientific explanation Michael Crichton embedded in the book. Indeed, by the figure of the cynical mathematician Ian Malcom, played by Jeff Goldblum in the movie, Crichton greatly popularized chaos theory among laypeople.

complex, may display some non-computable statuses, but their dynamics are clear and can be actually modeled as far as they are governed by a series of *physical* causations. As soon as we move into the biological and then psycho-cognitive plan, we are not dealing with causations but with much looser (and hence uncertain) forms determined by the interactions between the two, or more, elements: we can speak of *enablements* [20] or, as the cognitive aspects are concerned, of *affordances* [11].

Many of the phenomena a cognizing organism has to cope with are much more similar to the second example (lion-gazelle) than to the first. Survival (encompassing both the notions of fitness and welfare) rests on the possibility of continuously appraising unpredictable situations and making the best judgment out of them.

Nevertheless, it is commonly agreed that higher cognitive capabilities are essentially about saving the cognitive effort, and one of the best ways to do this is to realize that unpredictability and similarity are not mutually exclusive and that, even if every biological (and then social) phenomenon is a priori unique, it is possible to elaborate certain heuristics in order to exploit the *approximate-cause-effect* relationships nested in biological unpredictability and randomness. This is what Tooby and De Vore described as *accessing* the "cognitive niche."

> At the core of this lies a causal or instrumental intelligence: the ability to *create and maintain cause-effect models* of the world as guides for prejudging which courses of action will lead to which results. Because there are an infinitely large number of possible sequences of behavior (of which the overwhelming majority are maladaptive) "behavioral flexibility" is an insufficient characterization of our innovative adaptive pattern. Our cognitive system is knowledge or information driven, and its models filter potential responses so that newly generated behavioral sequences are appropriate to bring about the desired end. Of course, exploration, trial and error, and feedback are essential to the system, but, by themselves, they are inadequate to construct or maintain it [41, p. 210, added emphasis].

If Tooby and De Vore (and subsequently Pinker [36]) understand the cognitive niche as a kind of *stage* of cognition, in their opinion exclusive to human beings, other scholars such as Clark [6] and Magnani [23, Chap. 6] have a much more local and objectified vision of a specific cognitive niche as something "constructed", which can be defined as follows.

> Cognitive niche construction is the process by which organisms modify their environment to affect their evolutionary fitness by introducing structures that facilitate (or sometimes impede) the persistent individuation, the modeling, and the creation of cause-effect relationships within some target domain or domains. These structures may combine with appropriate culturally transmitted practices to enhance problem-solving, and (in the most dramatic cases) they afford potential whole new forms of thought and reason [2].

Along the next sections we will better define the notion of cognitive niche (in its "constructed" connotation), and see how it can account for the way human "cultural" traits (ranging from hunting to advanced economy) can be seen as increasing welfare by reducing unpredictability, but also how such activities may hinder survival by introducing a new, subtler level of uncertainty.

2 Cognitive Niche Construction: Managing Resources to Contrast Uncertainty and Lessen Complexity

In a nutshell, a cognitive niche consists in a series of externalizations of knowledge into the environment, for instance through material culture, resulting in a modification of the selective pressure that an organism has to face [23, 32]. The fact of championing cognitive niche construction could be seen as what intrinsically characterizes human beings (which are individuated by the theory as *eco-cognitive engineers*). The rest of the paper will then focus on the notion of terminator niche: a cognitive niche that becomes maladaptive *because* of the externalized knowledge structures that primarily did (or were thought to) cause the beneficial trade-off in selective pressure.

Since the dawn of cognition, we have been acting on our surrounding ecologies in order to make them easier to live in, and we engineered niches ranging from basilar sociality [10] to material culture [29], through agricultural and hunting abilities. Every single step of development can be framed within the concept of eco-cognitive engineering: we engineer our environment by externalizing and manipulating pieces of knowledge. Otherwise said, humans (like other creatures) do not simply live *in* their environment, but they actively shape and change it while looking for suitable chances. In doing so, they construct *cognitive niches* through which what the environment offers in terms of cognitive possibilities is appropriately selected and/or manufactured to enhance their fitness as chance seekers [35, 36, 41]. Lessening the selective pressure means, for our cognitive efforts, to lessen the complexity of the external world by developing simpler models of how the environment works, and to enact them making the world a less unpredictable place to live in.

A recent book by Odling-Smee et al. [32] offers a full analysis of the concept of niche construction from a biological and evolutionary perspective. "Niche construction should be regarded, after natural selection, as a second major participant in evolution. […] Niche construction is a potent evolutionary agent because it introduces feedback into the evolutionary dynamics" [32, p. 2]. By modifying their environment and by their affecting, and partly controlling, some of the energy and matter flows in their ecosystems, organisms (not only humans) are able to modify some of the natural selection pressure present in their local selective environments, as well as in the selective environments of other organisms. This happens particularly when the same environmental changes are sufficiently recurrent throughout generations and selective change.

In summary, general inheritance (natural selection among organisms influencing which individuals will survive to pass their genes on to the next generation) is usually regarded as the only inheritance system to play a fundamental role in biological evolution; nevertheless, where niche construction plays a role in various generations, this introduces a second general inheritance system (also called *ecological inheritance* by Odling-Smee). In the life of organisms, the first system occurs as a one-time, unique endowment through the process of reproduction (sexual for example); on the contrary, the second system can in principle be performed by any organism towards any other organism ("ecological" but not necessarily "genetic" relatives), at any stage of their lifetime. Organisms adapt to their environments but also adapt to environments as reconstructed by themselves or other organisms.[3] From this perspective, acquired characteristics can play a role in the evolutionary process, even if in a non-Lamarckian way, through their influence on selective environments via cognitive niche construction. Phenotypes construct niches, which then can become new sources of natural selection, possibly responsible for modifying their own genes through ecological inheritance feedback (in this sense phenotypes are not merely the "vehicles" of their genes).

It has to be noted that cultural niche construction alters selection not only at the genetic level, but also at the ontogenetic and cultural levels. For example, the construction of various artifacts challenges human.

> Humans may respond to this novel selection pressure either through cultural evolution, for instance, by constructing hospitals, medicine, and vaccines, or at the ontogenetic level, by developing antibodies that confer some immunity, or through biological evolution, with the selection of resistant genotypes. As cultural niche construction typically offers a more immediate solution to new challenges, we anticipate that cultural niche construction will usually favor further counteractive cultural niche construction, rather than genetic change [32, p. 261].

With a broader explanatory reach than sociobiology and evolutionary psychology, the theory of niche construction simultaneously explains the role of cultural aspects (transmitted ideas), behavior, and ecologically persistence inheritance. Of course niche construction may also depend on learning. It is interesting to note that several species, many vertebrates for example, have evolved a capacity to learn from other individuals and to transmit this knowledge, thereby activating a kind of proto-cultural process which also affects niche construction skills: it seems that in hominids this kind of cultural transmission of acquired niche-constructing traits was ubiquitous, and this explains their success in building, maintaining, and transmitting the various cognitive niches in terms of systems of coalition enforcement. "This demonstrates how cultural processes are not just a product of human genetic evolution, but also a cause of human genetic evolution" [32, p. 27]. From this viewpoint the notion of *docility* [38] acquires an explanatory role in describing the way human beings manage ecological and social resources for decision-making.

---

[3]This perspective has generated some controversies, since the extent to which modifications count as niche-construction—thus entering the evolutionary scene—is not clear. The main objection regards how far individual or even collective actions can really have ecological effects, whether they are integrated or merely aggregated changes. On this point, see [40] and the more critical view held by [8]. For a reply to these objections, see [18].

2.1 Economy, Economics and Cognitive Niche Construction

Finally considering the economy, it can be rightfully seen as relating to cognitive niches from its very etymology. The word derives its current 17th century meaning from the Greek *"oikonomia,"* household management, based on *"oikos,"* house + *"nemein,"* manage. The ancient notion of *household* incorporates the idea of cognitive niche, indicating at the same time the ecological diffusion of the family (the physical estate), and the maintained set of rules (shared to different extents) that regulate it. Speaking of household management epitomizes the idea of cognitive niche, by resuming all of the heuristics and problem-solving activities aimed at facing the cause of problems, that is the unpredictable conjunctures of the "external" world, upon which all households would survive.

It is easy to extend the discourse of households to the actual economic discourse including firms, stakeholders and other actors. In its more actual significance, economic systems can still be seen as a broad cognitive niche, where knowledge representing cause-effect patterns in the unpredictable and complex economic world are individuated, in order for the actors to manage the available resources pursuing a determinate goal. Indeed, economics, as a discipline and a practice, consist in a niche construction and regulation activity supposedly making sure that the adopted strategies are coherent with the state of resources and with the goals they aim at producing. Indeed, as Dow observed, "the emergence of institutions more generally can be understood as a means of reducing uncertainty […]. The existence of the firm itself (indeed, of any contractual arrangement) can be understood as a mechanism to reduce uncertainty for some parties involved, providing a pool of liquidity and a basis for action" [9, p. 40].

Several perspectives on uncertainty in economics have already approached the matter in a way that is compatible with cognitive niche construction theories: consider the instructionist approach focusing on firms, claiming that the latter avoid uncertainty by restricting to ecological niches that are simple and change very slowly [1, 34], and of course the evolutionary approach[4] [31]. Therefore, assuming that the description so far satisfies the connection between economics and cognitive niche construction theories, what we aim at analyzing are critical economic conditions through the latter theory: in order to do that, we will first briefly spell out key features of high-technology cognitive niches (of which contemporary finance is a clear example), and then analyze the latest economic crisis in the light of the developed argument.

[4]The evolutionary approach usually focuses on micro-economics, but one of the values of cognitive niche theory is that cognitive niches can be overlapping, coextensive and included one into the other—think of some major, foundational niches such as language [6].

Before moving on, the fundamental issue of open and closed systems has to be brought to attention once again, to understand the full impact of cognitive niche construction for the economic discourse. As shown by the already mentioned Cecile Dow, the understanding of uncertainty and its impact on the economic practice is a largely epistemological issue [9]. The economists' assumption and the behaviors they prescribe depend on their beliefs concerning the capability of cognition to cope with uncertainty and unpredictability. As a powerful cognitive niche, the economy *and* economics play a crucial role in improving the profitability of a system by reducing and computing aspects otherwise contemplated as non-computable. At the same time, it must be remembered that the construction of a cognitive niche takes place *over* a given system, as a form of scaffolding, to rely on Clark's most fitting metaphor [6].

This provides niche users with a closed, man-made system that allows better predictions and better control over an open, highly uncertain system. Nevertheless, this does not imply that niche construction delivers a 1:1 control over the original system. Consider agriculture, which clearly instantiates a good idea of cognitive niche construction: the invention of agriculture turned the environment into a much more reliable provider of food through a vast amount of shared knowledge and heuristics, but this did not guarantee the end of famines and the optimal yield in every crop forevermore. Economics share the main goal of cognitive niche construction, which is to improve cognitive capabilities by reducing unpredictability and uncertainty: because of the shift between the niche-system and what the niche is constructed upon, assuming that economics (*qua* cognitive niche construction) may utterly remove elements of uncertainty is a powerful and dangerous misunderstanding.

Such misconception, we will argue, can be fostered by particular structures embedded in the niche. In the economic discourse, the problem relies in the mainstream assumption that "uncertainty represents a lack (of certainty or certainty-equivalence) which prevents agents from fully informed rational optimisation, so that it is seen as anathema (particularly in financial markets)" [9, p. 43].

It should also be kept in mind that—given its strong link to a traditional idea of logicality, in which the need of formalization and quantification dominates—economic language appears to be compelled to ineluctably depicting unpredictability as an effect of "uncertainty". In this perspective concepts like preference, utility, welfare or uncertainty are at the core of economic mainstream analysis, just because they need to be quantified. A wider attention to recent logical developments could help to favor a relaxed but more fruitful consideration of the problem of uncertainty: new non standard logical perspectives, which are no more endowed, like in the case of classical logic, with universal characteristics, could offer a better logical-epistemological framework for economic cognition. For example, the current interest in the so called "naturalization of logic" [25, 26, 42] and in abductive reasoning certainly stresses the urgent need of abandoning a global *arithmomorphic* epistemological attitude reverting to a more "anthropomorphic" one, in which logical modeling is closer to human local cognitive ways of producing inferences and argumentations in the so-called "eco-cognitive perspective."[5]

2.2 *Virtualized and High-Tech Cognitive Niches*

The theory of cognitive niches is extremely valuable because it allows not only to understand human cultural development in its traditional meaning, but its frame can be extended to comprehend hyper-technological cultures as well: the situation we sketched out so far could be say to work at least until the Fifties of past century. Then something changed: until then, a cognitive niche could be described as a relationship between *biota*, *abiota* and dead organic matter. Either you are alive, and then you can be a constructor, or you are not, and then you are a constructed. What is constructible is the *object* of cognitive niche construction: it is the target and the materiel on which the externalization of knowledge was built. And that was it. Since the computational revolution, though, cognitive niche construction was enhanced by something that was neither a biota, nor an abiota or dead organic matter: it was the category of *constructed constructors*.

The most important breakthrough of high-tech niche construction involves the production of more or less complicated "artificial minds" [21]. The notion of artificial mind can be seen as an help, or as a "maid-mind," but the aim is the same, that is to obtain a new kind of eco-cognitive engineer that contributes to the activity of niche-construction.

Virtual niches, and high-technological niches, are populated by a number of *constructed constructors*, that is by agencies that were constructed (or programmed) externalizing knowledge on abiota materiel, but can actively engage a more or less extended range of active behaviors within the niche. These new actors can either chiefly serve either as assessors, maintainers and mediators of existing externalizations, or as engineers of new externalizing solutions in the niche, or as full-right agents in the cognitive niche.

These actors need not be "material:" those interacting within traditional cognitive niches (such as driving supporting systems) tend to be material, but they can also reside in a bit of coding, such as a data mining software, and yet be able of

[5]In human reasoning, especially of the premise-conclusion sort, it is comparatively rare that the standards of deductive validity or statistico-experimental inductive strength are actually met. Accordingly, new standards for the so-called 'third way' reasoning [42, Chap. 7] should be found as a mixture of nonmonotonic, default, *ceteris paribus*, agenda-relevant, inconsistency-adaptive, abductive reasonings, for which neither standard deductive validity nor inductive strength are a good rule of assessment. The best recourse involves a significant restructuring of the varying provisions of families of nonmonotonic logics, crucially including the logic of abduction. Most 'right' reasoning is third-way reasoning, which owes its rightness to the meeting of requirements other than deductive validity (or inductive strength) [25].

causing significative modification to the global structure of the niche. In all of these cases, the crucial feature is the presence of non-human cognitive agents, usually embedded within a cognitive niche, that are able to:

- Assess a situation.
- Make an appraisal.
- Take a decision based on that appraisal.

The final decision, which is usually the contribution to the cognitive nice (for instance in the shape of an affordance) is meant to be for the good of the human user —or at least of some human users, as in the case of "intelligent" weaponry [17]. As we stated several times in this subsection, the revolutionary steps consisted in the assumption of non-biological material to the status of actor in a cognitive niche: it is not the same as stating that, for the first time, the new status was given to something different than a human being: animals have traditionally been actors of cognitive niches, also as assessors and decision makers (a trivial example: watchdogs are expected to be able to tell a friend from a foe), but animals are part of the biota, they are trained and not constructed, and do sometimes actively resist niche construction activity. Conversely, in high-tech cognitive niches new actors are introduced, and they are shaped precisely as their creators want them to be.

Another relevant feature of high-tech cognitive niches is the presence of *cyborgs* [5, 22]. This is not the place for a discussion of cyborgs, but they are worth mentioning because not only we witness the delegation of cognitive niche construction to artificial agency, but also biological agents, the traditional constructors, are further and further hybridized with the technological artifacts, so that the limit situation could be described as a combination of automatic niche construction activity and cyborg niche construction activity. In other words, the high-tech cognitive niche could be seen as supporting artificial decision maker and hybridized (part biota and part abiota) decision maker.

3 Economic Terminator Niches

What happens, though, when artificial-minds *as* eco-cognitive engineers cease to collaborate with human beings in the management of resources, for instance? Actually, the question is not accurate, since it would mean to imbue them not only with passive moral rights, but also with an intentional moral will: more properly, could it happen that such agents *keep pursuing the tasks they were endowed with by their human programmers in a way that is not beneficial to human beings anymore?* In order to answer this question, we must not forget that the essence of niche construction is in fact to lessen selective pressure, not to increase it making life more difficult or simply unsustainable: terminator niches need not be necessarily high-technology niches. In fact, the conditions for the emergence of a terminator niche are simple: the niche must turn maladaptive because of some of the structures that chiefly achieved (or were thought to achieve) the ease in selective pressure; and, the more the conditions caused by the cognitive niches grow sever, the harder it gets to revert and dismantle the cognitive niche.[6] Within an hyper-technological niche, as we will see, the terminator phase can acquire some peculiar characteristics —that depend on the discussion we just sketched out about high and hyper technological niches—but cognitive niches have already happened to turn the change of selective pressure *against* the human beings who had engineered them.

3.1 In History

One should not be lead astray by the label of "terminator niche." We are not (necessarily) thinking of androids chasing human beings, of machines rebelling and such things. In fact, terminator niches have been developed since the dawn of humankind. Consider this very fitting example provided by paleoanthropologist Steven Mithen. It is about the Natufian culture, which existed in Eastern Mediterranean from 13.000 to 9.800 years ago, and their way of managing hunting —which can be seen as an "economic" will to overcome the complexity of their world. Deciding to hunt a kind of prey rather than another is quite similar to decide to invest in financial operations rather than in opening a new production plant.

> When the Kebaran people had used the Hayonim Cave, five thousand years before the Natufian became established, they killed male and female gazelles in equal proportion. *By preferentially selecting the males, the Natufians were probably attempting to conserve the gazelle populations*. Although both sexes were born in equal proportions, only a few male animals were actually needed to maintain the herds. Carol Cope thinks that t*he Natufian people decided that the mass were expendable* while recognizing the need to ensure that as many females as possible gave birth to young.If this was their *aim*, it went *horribly wrong*. The Natufians made the mistake of not just hunting the males, but selecting the biggest that they could find to kill. So the female gazelles were left to reed with the smaller males—unlikely to have been their natural choice. As small fathers give rise to small offspring, and as the Natufians killed the largest offspring, the gazelles reduced in size with each generation. […] Smaller gazelles meant that there was lees meat available to feed an ever-growing population. This shortage was compounded by over-exploitation of the 'wild gardens': too many stalks of the wild cereals had been cute and excessive quantities of acorn and almonds had been collected for natural replenishment to occur. The health of the Natufian people began to suffer, especially that of the children. […] Food shortages can also lead to poor physical growth [30, pp. 47–48, added emphasis].

The final result of this process was emigration and the eventual abandonment of Natufian settlements. The point of this historical example is quite clear: not everything that *goes horribly wrong* can be the sign of the development of a

[6]The notion of *Terminator Niche* is obviously modeled upon 1984 sci-fi action blockbuster *The Terminator*, in which Skynet, an AI system originally engineered to protect human beings from nuclear warfare, perceives humans as a threat and attacks first, causing nuclear holocaust and then building androids ("Terminators") to hunt down remaining humans.

terminator niche. For instance, natural mishaps can take place. Climate changes before they were linked to human activity. Human beings can develop strategies that are ultimately harmful for themselves. Yet, we do not think that *smoking*, for instance, can be labeled as a terminator niche. Why? Because smoking, in a strict sense, was not developed as a mean to reduce selective pressure. It is an intrinsic activity, carried out for its own sake because it is found pleasurable. Conversely, the Natufians' situation can be resumed as follows:

1. *Hunting* is a cognitive niche because it can be described as a set of heuristics, techniques and affordances aimed at providing better food income than if preys were chased randomly.
2. Hunting heuristics favor the intentional killing of larger males as this seems to provide two advantages:
    - More food is purveyed to the group.
    - Herds are maintained since the male to female ratio can be very small and yet permit the numerical prosperity of the herd.
3. The hunting heuristics, on the long term, induces a diminution in the size of the prey, hence in the quantity of food purveyed to the group: this is the opposite of the original strategy.
4. The survival of the group is ultimately jeopardized by the cognitive niche it structured.

A characteristic of cognitive niches is that they have to be maintained in order to function, hence there is a *conservative* drive towards making externalizations persistent and resilient. This accounts for the fact that many human achievements are hard to eradicate even when they prove less than ideal. Cognitive niches resist being easily wiped away because their original scope is to contrast selective pressure. Externalizations must be therefore persistent to oppose exogenous changes. This, of course, makes things harder when the increased weight of selective pressure is happening *because of* the niche that should shield us from it.

In the case of the Natufians, the niche goes *terminator* because of the interplay between stages (2) and (3). This spells out a very interesting character of terminator niches. It is possible to individuate a *point of no-return*. The Natufian crisis did not happen overnight, so some individuals could—at a certain point—individuate the negative trend, relate it to the hunting practice, and call for it to stop. We do not know if this happened among the Natufians. Historically, even when such enlightened individuals made their appearance, they often went unheard.[7] Yet, after the stage where the negative trend is apparent to some, the *terminator niche* goes in full bloom at the moment when any solution is not, or seems not, viable anymore. Imagine that at time t^1 there were no "big" male gazelles anymore, but there were

[7]This aspect will be examined further on about the emergence of contemporary finance as a terminator niche. As cognitive niches acquire also a moral value as orthodoxies, we can expect a violent reaction against those who suggest safer alternatives just because they are alternatives [24].

medium-sized one, small and smaller ones left. At this time, it could be sensible to refrain from preying on the medium sized ones, and concentrate on smaller ones: if the niche had developed and incorporated such heuristic, it would not have turned terminator. The situation at t^2 is different: you do not have any medium sized gazelles left, you only have small, and smaller, so you have to harvest empty most of the spontaneous natural cereal cultures. At this point, there is no viable solution left: when small gazelles are not enough to sustain the group, it makes little sense to shift to hunting even smaller ones. The situation could only degenerate, and human beings tend to dislike accelerating their doom, however unavoidable this seems to be. As history tells us, the Natufians' terminator niche culminated in the abandonment of the niche itself, with the emigration of the population.

3.2 *Neo-Liberal Finance as a Terminator Niche*

As suggested in recent research and reviews about uncertainty in economics,

> Much of the standard mainstream economics and finance literature ignores uncertainty by conflating it with quantifiable risk. Even though it may be accepted that risk cannot be quantified in general in objective terms, nevertheless it is argued, according to the subjective expected utility model, that we have the capacity to make subjective probability estimates, so that unquantifiable risk is no longer relevant [9, p. 34].

Such conflation, as we will see, is about a partly willing theoretical and epistemological commitment that drives the economic niche construction. Economists, acting as gatekeepers in niche construction activities, transmit their epistemological status about uncertainty to the economic system, under the tacit assumption that the functioning of the economy is basically a self-realizing prophecy not only for the bad but also for the good. Reducing uncertainty to quantifiable risk by means of shared beliefs, narratives, heuristics and best-practices should indeed make such reduction happen. The reason why such process is so subtle and insidious is that for a good deal of time, precisely because of how the niche is structured, it works—at the same time setting the stage for the catastrophe of the same system: "indeed, it was financial stability which bred instability" (p. 37). Stapleton, a few years earlier, provided a reconstruction of the latest financial crisis highlighting the role of the illusions embedded in the system by means of shared assumptions (diversification warrants risk obliteration) and the role of the communicative and computational structure managing the access to data and results.

> In the early 2000s, financiers believed that, through our integrated financial systems, we could fragment and disperse loan risk so much as to make that risk completely negligible. Risk itself would magically disappear in the ecstasy of post-structuralist communications […]. Like a starship, financial risk would at last achieve escape velocity and reach the financial galactic beyond. And so we created the giant Ponzi scheme known as the international financial system based on almost infinite hedging and fund fragmentation and dispersal, all made possible by our integrated global financial technologies. Like

pre-enlightenment financial alchemists, we could turn base sub-prime loans into gold. Instead, we found that we turned it into a global bank debt crisis and eventually a sovereign catastrophe [39, p. 5].

Stapleton's analysis is unforgiving. Or even better, it is "curiously" forgiving inasmuch as he does not approach the crisis from the financial point of view, but from that of hyper-technological cognitive niches. Focusing on the crack of the Anglo-Irish Bank, he claims that the fault is not to be found in masterminds of crime or "slackerism," but rather in the decision-making system that was cyborg-like, shared between humans and the computers they had—themselves—programmed. "The justification for action is expressed in the form of a 'conviction narrative' that presents an argument which is convincing in that it is in itself coherent" [9, p. 39].

What was the role of management information systems in all this? It was surely these systems that facilitated financial imprudence and light tough regulation, simultaneously providing a sense of a controlled and well-monitored business. Rather than deliver solid management information to support wise decision-making processes, the systems not only failed, but created an illusion that all was well. [...] Thus, management does not gain a real-time, true and integrated picture of their firm. Instead, technology and culture operating together in this Faustian tryst produce the very opposite effect: an illusion of prudence and effective risk management. A technoculture of deceit, of hiding and cover-ups, is therefore potentially enabled by our technology-cultural system [39, p. 6].

Dow reaches a similar conclusion about the epistemological and inferential responsibilities adding up to the outbursts of the financial crisis, stressing the role of the *institutionalized modeling approach* at play in the economic mainstream:

But the financial crisis arguably stemmed from mechanisms that fostered over-confidence in expectations as to risk and return, that is, an inappropriate inattention to uncertainty. [...] A powerful conflict of narratives is provided by the fact that financial markets have relied increasingly on quantitative models which exclude uncertainty, limiting scope for judgement, whilst the experience of uncertainty became palpable in the crisis. [...] Strategy in the financial sector in the run-up to the crisis arguably was shaped by institutional narratives that were unduly influenced by the excessive confidence that arose from a basic modelling approach that ignored uncertainty. Indeed, this modelling approach was institutionalised by the very capital adequacy requirements which were intended to reduce risk. The mechanism that therefore evolved in financial markets to address uncertainty in the run-up to the crisis was denial [9, pp. 40–41].

Can finance be defined as a terminator hyper-technological niche? We believe that the categorization is fitting.

1. With the benefit of a powerful charity principle, we can say that finance is a cognitive niche constructed in order to increase total welfare, albeit in a capitalistic conception of markets.
2. Finance is a high-tech cognitive niche, as it is greatly virtualized and its actors are not only human beings but software and algorithms (such as those for risk-assessment) and other forms of robotic intelligence: automatic trading systems are common, whereby trades are triggered automatically in fractions of a second when some asset price reaches a pre-specified level. This, in turn, means that human agents involved are significantly cyborg*ized*.

3. Karl Marx had already theorized that crises are endemic to the structure of capitalism, but crises following speculative bubbles such as the 1929 one that spurred the Great Depression, and the 2008 subprime loans one that turned into the ongoing global crisis, seem to be shifting from something structural to something that is jeopardizing the welfare (and potentially the survival) of those who populate and maintain the niche.

Albeit they did not use the term "terminator niche," since they are not adopting the niche theory at all, many economists (for instance in the PostKeynesian school) have argued about the intrinsic unstable nature of financial markets (seminal work of Minsky) and have stressed the pervasiveness and the disruptive nature of an excessively *financialized* economic system. Neoclassical finance considers economic agents as entirely rational (this trust was extended to the hybrid and artificial agents of the past few decades), and—basing on this philosophically uncertain assumption—developed models that too often mix up "risk," as something that can be *measured*, and Keynes' concept of fundamental "immeasurable uncertainty."[8]

Minsky, in 1963, claimed that financial markets are intrinsically unstable because of debt structured built by economic agents (namely Ponzi schemes), that will sooner or later cause the collapse of the whole system [28].

More specifically, he argued that stability breeds instability since confidence in rising asset prices encourages ever-more leveraging (the Ponzi structure just being the most extreme), increasing vulnerability to asset price reversals. The key role for uncertainty (an absence of confidence[9])—it is shifting levels of confidence in price valuations (due to some development or other—in this case, realization of subprime mortgage default) which brings on the crisis. This instability has been an ongoing phenomenon, accounting for periodic crises. but the increasing role of mathematical modeling in finance could be said to have caused the extreme instability of the recent crisis.

[8]For specific events (for instance a roulette table) we can calculate the probability of the outcome. Conversely for others—such as catastrophes and other events, which have been often used as the underlying of many derivatives instruments—we just cannot measure the probability of the outcome. "Amongst financial institutions themselves, financial products such as derivatives and credit default swaps were developed to reduce uncertainty, although in aggregate the effect turned out to be the increase in uncertainty on the onset of crisis" [9, p. 40]. Indeed John Maynard Keynes himself, in his "Treatise on probability" [14], neglected the possibility to build up a theory of expectations, even with help of probability calculus.

[9]Here understood as a pragmatic result. It depends on how you define uncertainty, an issue reinforcing the plea for a deeper epistemological analysis of economics. If we use the Knightian definition of uncertainty [16] uncertainty is an "estimate": a situation where the decision maker does not know the frequency distribution of the outcomes of a series of instances either because these instances are not homogeneous hence they cannot be grouped, or because historical data are not useful to predict the future. Often subjectivist theorists probability corresponds to the so-called "degree of belief" in a given proposition or event. (e.g. Savage, Lucas) have interpreted Knightian uncertainty as a lack of confidence in the likelihood of the occurrence of an event.

> Minsky's Keynesian theory of financial instability sets out the process by which conventional expectations become more confidently held during an upswing (although the conditions for uncertainty continue to be present), such that planned investment increases and finance is more readily available, reinforcing this confidence. The multiplier-accelerator effects of this increased investment fuel economic expansion, which lends further confidence to investment planning and its associated finance [9, p. 37].

With the same respect, it should be remembered that when the latest economic crisis was far from exploding, Structural Keynesian economist James Crotty showed that:

> NFCs [US large Nonfinancial Corporations] were eventually placed in a *neoliberal paradox:* intense product market competition made it impossible for most NFCs to achieve high earnings most of the time, but financial markets demanded that NFCs generate ever-increasing earnings and ever-increasing payout ratios to financial agents or face falling stock prices and the threat of hostile takeover [7, p. 1].

For the sake of brevity, we have to make very short a story that would be much longer. We have a cognitive niche (finance) which impose itself over market competition, but which cannot make the necessary gains from market competition (which conversely it impairs), therefore it creates some proper schemes for increasing its welfare by assuming counterintuitive principles such as the rationality of economic agents and the illusion of control by calculating risk through unrealistic mathematical models. It is not necessarily to postulate *evil*, this is how cognitive niches work: furthermore, as argued by Stapleton, the reliance on an artifactual hyper-technological niche blissfully blinded (and still blinds) many operators: finance in many cases is not a mere self-fulfilling prophecy, but a prophecy that aims at being self-fulfilling, but falls short of it because prophets are not even humans but cyborgs or computational intelligences. This whole mechanism rings a bell, but where did we see it... Ah! Of course, the good old Natufians! Actually, the description of finance we just sketched out, which is quite an approximation but consistent with reliable economic analyses, is not that different from adopting the hunting decision to kill the biggest male gazelles, so to get more food *and* let the population of preys thrive by not subtracting females to the herds. Too bad history proved the Natufians wrong: human beings have an innate desire to have their cake *and* eat it. A certain kind of terminator niche can be seen as the externalization of this desire. In particular, hyper-technological cognitive niches can make the actualization of this desire as something more possible, and at least at the beginning they make it happen: computational intelligences, if "properly" programmed, can create whole systems of meaning and whole possibilities of action that appear as viable, albeit in traditional cognitive niches they would be quickly debunked as unfeasible.

Finance as a terminator niche embeds the perception of ineluctability that also emerge from the prehistorical example of the Natufians: as everybody witnessed, current politics, aimed at *regulating* markets, coupled with generous insertions of liquidity from the Federal Reserve and European Central Bank, have not achieved a stable recovery yet. According to economists such as Palley, the only way out would be to revert the financialization of the entire world economy [33].[10] But asking this seems like telling a hungry Natufian to eat an even smaller gazelle so that the situation *might* improve.

Because human being are constantly constrained by the need to economize on mental resources, cognitive niches are plagued by the desperate need to believe what is commonly said [38]. This, summed to a tendency towards resilience and persistence that is vital for the maintenance of cognitive niches, triggers a sclerotization of terminator niches [24, Chaps. 4 and 5]: the more they fail in offering a positive trade-off in selective pressure, the harder human beings cling to them. Each time the current financial crisis seems to be touching an all-time low, neoliberal think-tanks (such as the Tea Party movement in the US) call for harsher neoliberal politics. Telling people "It has worked till now, it will recover and work again"—notwithstanding the epistemic scarcity of inductive reasoning—is more welcome than alternatives such as "This is not working anymore, we have to look somewhere else for a solution."

When considering the reaction to the financial crisis in a cognitive niche perspective, it is important to see how even power relations can be framed within the theory. Indeed, power relations are part of a *coalition enforcement* scheme, where a coalition is seen as a group maintaining the cognitive niche it lives in [3, 24, 37]. When speaking of "mainstream" economics/economists, we referred to them as gatekeepers regulating the good practices of niche functioning. As seen in the long (and arguably achieved) aftermath of the 2008–2010 crisis, the governments seemed to appease the economic free fall by increasing deregulation (and hence precariousness), enforcing austerity (reducing state expenses and state-expenses-fueled consumptions), and assuming a much harder stance against public and national deviants with respect to private ones (favoring private sector bailouts but implementing dire consequences for national ones, as in the case of Greece). This indeed is not about *evil*, it is about the doxastic inclination towards self-preservation of the cognitive niche, especially in cases in which a cognitive niche such as the financial sector is overflowing into the other domains of the life of a nation.

After all, it is not a mystery that after the Second World War a fundamental equivalence was instituted between defending the territorial integrity of the Western World, defending its political systems, defending and exporting the "American way of life" and the *defense of capitalism*: as that mindset essentially shaped the world we live in now, and most of our comforts, we tend to obliterate the *contingent* nature of recent economic/political history in favor of a vision of *necessity*. Still, things are indeed grimmer when the dominating niche supporting power relations is in a *terminator* phase. In popular culture, terminator niches are characterized by a progressive loss of rational hope, as the everyday scenario turns grimmer and more

[10]We are talking about dismantling a cognitive niche. History shows that, in order to break the resilience of a cognitive niche, significant impetus is required: for instance, massive invasions, cataclysms and similar things.

distopian. This general depression goes hand in hand with religious and mystical hope, prophecies and *deus ex machina* solutions (movies such as The Terminator, Matrix and so on are perfect examples) but also with human beings getting violently carried away by a sacrificial state of mind: as spelled out by Girard in *The Scapegoat* [12], crises spark victimary mechanism in which random victims, fittingly individuated, are blamed for the situation and punished, expelled from the community or killed, hoping that this will bring relief.

This theme, ever-present in sci-fi terminator niches, could be traceable also in finance-as-a-terminator-niche: after all, if the cause of the actual crisis is the neoliberal finance-dominated economy, the perseverance in blaming as culprits whole nations (such as Greece, or Spain, or the next in line) or the occasional crook or rogue trader is nothing but the spectral appearance of a sacrificial mentality. Monetary bloodshed—fostering poverty, and degrading the potential of whole nations—may dope the terminator financial niche into believing that *hope* and *optimism* (ironic words for a system believing in the rationality of all agents and in the *unreal* calculability of any kind of risk) are finally justified. I can hope to have my cake and get to eat it at the same time, and this can make me merry for the whole day, and during the day I can celebrate with a shopping spree: this is the kind of illusion that finance, as a terminator niche, fosters. But come evening I will discover the bitter truth: if I ate my cake, I do not have it anymore and on the morrow I will be cake-less and miserable.

4 Conclusions: Crises as the Unpredictable Effects of Contrasting Unpredictability

In this paper, we first sketched out a brief theory of cognitive niche construction as a human artifactual effort aimed at distributing knowledge in various forms in order to cope with the unpredictability of the environment: a definition that would well encompass economic systems as we know them. Then, we introduced actual risks connected to the technological delegation of cognitive tasks to the niche (which should otherwise have a merely supporting role) and finally showed how finance, as a high-technology niche, could see its crises explained by cognitive niche degeneration, especially in the appearance of "terminator" cognitive niches.

Before concluding, we must stress that our contention is not an anticapitalistic one, nor does it share any common ground with Luddism. Not all the past ways of trying to harness uncertainty into unpredictability, and consequently into computable risk, were necessarily doomed and dooming. Consider for instance the Black–Scholes–Merton model, a mathematical model of a financial market containing certain derivative investment instruments [4, 27]. From the model, one could deduce the Black–Scholes formula, which gives a theoretical estimate of the price of European-style options. The formula led to a boom in options trading and scientifically legitimized the activities of the Chicago Board Options Exchange and other options markets around the world. It was widely used, although often with adjustments and corrections, by options market participants. Many empirical tests have shown that the Black–Scholes price was *fairly close* to the observed prices, although there are well-known discrepancies such as the "option smile", in "stable contexts". The Black–Scholes–Merton model is a counter example in that it only takes account of risk and not uncertainty, as separated by [16], and managed to work fairly well until the 1990s: a hedge fund administered by (among others) Scholes and Merton, the famous Long-Term Capital Management L. P., was initially successful with annualized return of over 21 % (after fees) in its first year, 43 % in the second year and 41 % in the third year, in 1998 it lost $4.6 billion in <4 months following the 1997 Asian financial crisis and 1998 Russian financial crisis, requiring financial intervention by the Federal Reserve, with the fund liquidating and dissolving in early 2000.[11] The BSM model is now known in the literature as over-simplified assumptions. Nonetheless before the crisis has been used by many practitioners as adequate to model option pricing [13].

Conclusively, let us recall a key feature that cognitive niche construction directly inherits from niche construction theories: the crucial relevance of feedback cycles. In a nutshell, the alteration of the niche (cognitive, or techno-cognitive in the case of finance) modifies the selective pressure. This new pressure should grant to the modifying agent a better management of uncertainties and a less complex landscape. Nevertheless the new, modified environment has a selective pressure that is *uncertainly* different both from the original one, and from the one of the desired environment. This introduces a second form of unpredictability, that is different from the one that the agent detects in the environment: indeed, this stronger unpredictability encompasses the niche constructors as well, rendering niche construction an effort towards curbing the unpredictability of the world, but which is characterized itself by a further unpredictable dimension. Niche construction theory can be of help in understanding the economic and financial system. The uncertainty is partly assumed, and most often the actual effects of massive speculations or central bank maneuvers are merely hypothesized. Sometimes, though, the whole niche (or partial sub-niches) acquire the *terminator* trait by hiding what is uncertain, unquantifiable, hypothetical and hence withdrawable. Forgetting that a given cognitive niche is an effect of something that humans produced means to take it as a given, and to fall prey to the deeply unpredictable effects of the original struggle against the unpredictability of the resources.

One should be able to take a whole step back and appreciate the economy (and economics) as a powerful yet self-scaffolding cognitive niche, embedded with automatic computational systems capable of further niche construction and gate-keeping activities. Such appreciation is linked to the awareness that this cognitive

[11]This extreme example shows how popular was the BSM model and related formula. Sholes and Merton, who shared the Nobel prize for "new methods to determine the value of derivatives," were appointed in the board of director of a hedge fund management firm. This also confirms our contention about economists playing a crucial role in gate-keeping the construction of the cognitive niche.

niche is just a helpful sheet of heuristics, computations, cause-effect relationships and predictions that is "trapped" between the uncertain and unpredictable world it is built upon, and the "animal spirits" governing financial markets, in Keynes' perfectly fitting imagery:

> Even apart from the instability due to speculation, there is the instability due to the characteristic of human nature that a large proportion of our positive activities depend on spontaneous optimism rather than mathematical expectations, whether moral or hedonistic or economic. Most, probably, of our decisions to do something positive, the full consequences of which will be drawn out over many days to come, can only be taken as the result of animal spirits—a spontaneous urge to action rather than inaction, and not as the outcome of a weighted average of quantitative benefits multiplied by quantitative probabilities [15, p. 161–162].

However, this renewed awareness of the essential uncertainty and unquantifiable unpredictability characterizing the cognitive niche and its economic agents is not to be perceived solely as a negatively constraining feature. As contented by Dow, "the financial crisis arguably stemmed from mechanisms that fostered over-confidence in expectations as to risk and return, that is, inattention to uncertainty. It was argued that the uncertainty-denial adopted by financial markets and by mainstream economists has ultimately been counterproductive, actually increasing uncertainty. Furthermore, within a closed-system approach, fundamental uncertainty can only enter as an exogenous distortion, seen in negative terms. But whilst uncertainty can be debilitating at times, it can be seen at other times as being the counterpart to creativity and emergence" [9, p. 45]. And ultimately, uncertainty is the *counterpart* to new and better niche construction practices, heuristics, and technologies that are able to make economic agents act in spite of uncertainty and not blindly but systematically negating it.

Acknowledgements We would like to express our gratitude towards Giovanna Magnani for her commitment and patience in explaining many details of economic theories and for making sure that our epistemological outlook made sense within an economic discourse. We also wish to thank the distinguished editors of the special issue where this article was first published for accepting that it appears in this collection as well.

References

1. Anderson, C.R., Paine, F.T.: Managerial perceptions of strategic behavior. Acad. Manag. J. **18**, 811–823 (1975)
2. Bertolotti, T., Magnani, L.: Theoretical considerations on cognitive niche construction. Synthese (2016). doi:10.1007/s11229-016-1165-2
3. Bingham, P.M.: Human evolution and human history: a complete theory. Evol. Anthropol. **9**, 248–257 (2000)
4. Black, F., Scholes, M.: The pricing of options and corporate liabilities. J. Polit. Econ. **81**(3), 637–654 (1973)
5. Clark, A.: Natural-Born Cyborgs. Minds, Technologies, and the Future of Human Intelligence. Oxford University Press, Oxford (2003)

6. Clark, A.: Word, niche and super-niche: how language makes minds matter more. Theoria **54**, 255–268 (2005)
7. Crotty, J: The neoliberal paradox: The impact of destructive product market competition and impatient finance on nonfinancial corporations in the neoliberal era. Research Brief 2003–5 of the PERI—Political Economy Research Institute, University of Massachusetts-Amherst (2003)
8. Dawkins, R.: Extended phenotype—but not extended. A reply to Laland, Turner and Jablonka. Biol. Philos. **19**, 377–397 (2004)
9. Dow, C.: Addressing uncertainty in economics and the economy. Camb. J. Econ. **39**, 33–47 (2015)
10. Dunbar, R.: Gossip in an evolutionary perspective. Rev. Gen. Psychol. **8**(2), 100–110 (2004)
11. Gibson, J.J.: The theory of affordances. In: Shaw, R.E., Bransford, J. (eds.) Perceiving, Acting and Knowing. Lawrence Erlbaum Associates, Hillsdale, JN (1977)
12. Girard, R.: The Scapegoat [1982]. Johns Hopkins University Press, Baltimore, MD (1986)
13. Hunzinger, C.B., Labuschagne, C.C.: Pricing a collateralized derivative trade with a funding value adjustment. J. Risk Financ. Manag. **8**(1), 17–42 (2015)
14. Keynes, J.M.: A Treatise on Probability. MacMillan, London (1921)
15. Keynes, J.M.: The General Theory of Employment, Interest and Money. MacMillan, London (1937)
16. Knight, F.H.: Risk, Uncertainty and Profit. Hart, Schaffner and Marx, New York (1921)
17. Krishnan, A.: Killer Robots: Legality and Ethicality of Autonomous Weapons. Ashgate, Burlington, VT (2009)
18. Laland, K.N., Odling-Smee, F.J., Feld-man, M.W.: On the breath and significance of niche construction: a reply to Griffiths, Okasha and Sterelny. Biol. Philos. **20**, 37–55 (2005)
19. Longo, G., Buiatti, M.: Randomness and multilevel interactions in biology. Theory Biosci. **132**, 139–158 (2013)
20. Longo, G., Montévil, M.: Biological Time, Symmetries and Singularities. Springer, Berlin/Heidelberg (2014)
21. Magnani, L.: Semiotic brains and artificial minds. How brains makeup material cognitive systems. In: Gudwin, R., Queiroz, J. (eds.) Semiotics and Intelligent Systems Development, pp 1–41. Idea Group Inc., Hershey, PA (2007a)
22. Magnani, L.: Morality in a Technological World. Knowledge as Duty. Cambridge University Press, Cambridge (2007)
23. Magnani, L.: Abductive Cognition: The Epistemological and Eco-Cognitive Dimensions of Hypothetical Reasoning. Springer, Berlin/Heidelberg (2009)
24. Magnani, L.: Understanding Violence. Morality, Religion, and Violence Intertwined: A Philosophical Stance. Springer, Berlin/Heidelberg (2011)
25. Magnani, L.: Naturalizing logic. Errors of reasoning vindicated: Logic reapproaches cognitive science. J. Appl. Logic **13**, 13–36 (2015)
26. Magnani, L.: Naturalizing the logic of abduction. Logic J. IGPL **24**, 639–654 (2016)
27. Merton, R.C.: Theory of rational option pricing. Bell J. Econ. Manag. Sci. **4**(1), 141–183 (1973)
28. Minsky, H.P.: Can 'it' happen again? In: Carson, D. (ed.) Banking and Monetary Studies. Homewood, Illinois (1963)
29. Mithen, S.: The Prehistory of the Mind. A Search for the Origins of Art, Religion, and Science. Thames and Hudson, London (1996)
30. Mithen, S.: After the Ice: A Global Human History, 20,000 5000 BC. Harvard University Press, Cambridge (MA) (2004)
31. Nooteboom, B.: Learning, discovery and collaboration. In: Nooteboom, B., Stam, E. (eds.) Micro-foundations for Innovation Policy, pp. 75–102. Amsterdam University Press (2008)
32. Odling-Smee, F.J., Laland, K.N., Feldman, M.W.: Niche Construction. The Neglected Process in Evolution. Princeton University Press, Princeton, NJ (2003)
33. Palley, T.I.: Financialization: What it is and why it matters. Working Paper No. 25, The Levy Institute and Economics for Democratic and Open Societies Washington D.C (2007)

34. Pich, M.T., Loch, C.H., De Meyer, A.: On uncertainty, ambiguity, and complexity in project management. Manag. Sci. **48**(8), 1008–1023 (2002)
35. Pinker, S.: How the Mind Works. W. W. Norton, New York (1997)
36. Pinker, S.: Language as an adaptation to the cognitive niche. In: Chris-tiansen, M.H., Kirby, S. (eds.) Language Evolution, pp. 16–37. Oxford University Press, Oxford (2003)
37. Rohwer, Y.: Hierarchy maintenance, coalition formation, and the origin of altruistic punishment. Philos. Sci. **74**, 802–812 (2007)
38. Simon, H.: Altruism and economics. Am. Econ. Rev. **83**(2), 156–161 (1993)
39. Stapleton, T.: Zarathustra and beyond: exploring culture and values online. AI Soc. **28**(1), 95–105 (2012)
40. Sterelny, K.: Made by each other: organisms and their environment. Biol. Philos. **20**, 21–36 (2005)
41. Tooby, J., DeVore, I.: The reconstruction of hominid behavioral evolution through strategic modeling. In: Kinzey, W.G. (ed.) Primate Models of Hominid Behavior, pp. 183–237. Suny Press, Albany (1987)
42. Woods, J.: Errors of Reasoning. Naturalizing the Logic of Inference. College Publications, London (2013)

The Sciences of Finance, Their Boundaries, Their Values

Alex Preda

Abstract Several approaches from the social and natural sciences take finance, and especially financial markets, as a domain of systematic inquiry. Historians of economic thought have discussed extensively the emergence and evolution of some major, competing paradigms within finance, focusing on differences in their methodological and theoretical assumptions, as well as on the ways in which they have achieved dominant positions in the academia. However, how do these paradigms see the problem of their own value, in relationship to the value of their field of study? In other words, how do they present finance as a set of phenomena worth studying, and what is valuable about studying them from a particular angle? I examine here in this respect five significant scientific approaches to finance: financial economics, market microstructure, behavioral finance, social studies of science, and econophysics. I show how they represent the study of financial markets as a valuable, systematic endeavor, and how they represent their own value in providing a distinctive approach to the study of finance. I distinguish between internalistic and externalistic claims to value among these approaches. Internalistic value claims make reference to data accuracy and to methodological adequacy, while externalistic claims make reference to investigating links between finance and other forms of social organization and institutions.

A little while ago I was invited to a workshop at a business school in Continental Europe. The topic was the value of finance, where finance meant not so much the complex web of financial institutions and practices (some intrinsic to state structures, some public, and some private), but financial markets.

The gathering I was part of was an academic one. Scholars of finance met to exchange their views on the topic of value. These views were (and are) justified with respect to a scientific approach of what financial markets are. Views on the

A. Preda (✉)
King's College London, London, UK
e-mail: alexandru.preda@kcl.ac.uk

E. Ippoliti and P. Chen (eds.), *Methods and Finance*,
Studies in Applied Philosophy, Epistemology and Rational Ethics 34,
DOI 10.1007/978-3-319-49872-0_9

value of finance cannot be separated from views on what constitutes a science of the market, because cognitive assumptions, theoretical propositions, and methodological approaches for investigating market transactions will necessarily inform evaluations of markets, and therefore judgments about the value of the latter.

Moreover, views about the value of finance expressed within the framework of a science of finance are confronted with the issue of their own value: what is valuable about a particular conceptual and methodological approach to financial markets, in relationship to how that approach sees the value of finance, and in relationship to alternative approaches?

When I attended the workshop about the value of finance, I was expecting that the debates are going to be focused on values in the above sense. And yet, the discussion took a utilitarian turn: financial markets have value because they provide liquidity to social institutions. Still, a function is not a value, and definitely not an ethical value. Moreover, an effect of financial transactions (providing liquidity) cannot be taken for the value of financial transactions. What are the value(s) of finance?

Admittedly, financial markets are very complex too and the syntagm "markets" actually refers to a variety of material settings, historically developed institutions, actors, and social practices. Nevertheless, it is used to designate, generally speaking, the institutions participating in transacting a variety of instruments related to ownership rights or claims upon streams of income.

Seen from this angle, these institutions bring together participants whose mutual obligations rest exclusively upon the commitment to participate in such exchanges, instead of being anchored in family or ethnic group obligations. This is an insight already formulated by Max Weber almost a century ago, when he noticed that markets are "originally consociations of persons who are not members of the same group" [46, p. 639]. Yet, this does not mean that (financial) markets are free of ethical obligations—quite the contrary. It is exactly because transactions are free from fraternal ethics that they have to rely on their own, internal ethics system [46, p. 636]. Since any ethics system presupposes not only behavioral norms and prescriptions, but also values (which legitimate such norms), this means that financial markets—and finance, more generally speaking—cannot be devoid of values. Values, while internal to markets, cannot be seen as completely irrelevant for the society at large. (Financial) Markets bring together strangers and as such create webs of mutual obligations which, in modern societies, play a key role in socialization and group formation [46, p. 636]. Consequently, the values internal to markets cannot be seen in isolation from the values relevant to the society at large.

Weber's argument resonates with more recent ones formulated by Amartya Sen, for whom the value of finance cannot be kept apart from relationships of trust and responsibility, and therefore from ethical components [43, p. 47]. While fiduciary relationships are seen as justifying profit maximization for the principal (e.g., shareholders), the maximization principle is subjected to restrictions with regard to its consequences, and thus is never free of ethical constraints [43, p. 54], which are not limited to legal constraints.

If the value of finance cannot be separated from ethics, and if behavior in financial markets (including trading) is not free of ethics, then the science(s) of finance cannot avoid questions of value, especially (but not only) if they operate with particular assumptions about the behavior of market actors.

The question about the value(s) of finance in a broader societal context is also of particular significance if we think that, in the wake of the crisis of 2007–2008, critical voices have been raised about it. One famous question which has stuck with commentators was that formulated by the Queen of England at the London School of Economics: "why did nobody notice the crisis?" [35]. Meant to interrogate the practical value of academic knowledge about finance, this topic also points at how a notion of value(s) is integrated into the science(s) of finance, not only with respect to how these sciences see the value of their field, but also with respect to how they see their own value. Representations of the own value, in their turn, cannot be treated as separate from the conceptual and methodological assumptions with which these sciences operate, as well as from the data they use in supporting their claims.

1 The Sciences of Finance and the Value of Finance

What, then, are the sciences of finance and how do their theoretical assumptions resonate, more or less explicitly, in assumptions, or in outright judgments, about the value of financial markets? In trying to answer this question, one should deliberately avoid the singular. For decades, academic investigations of financial markets have been more or less the sole domain of financial economics. This situation, however, has substantially changed since at least the mid-1990s. Within the (social) sciences, we encounter nowadays not one, but several approaches to financial markets. These approaches have become institutionalized, a process manifest in formal educational programs, research agendas, publications, conferences, and recruitment processes in the financial services industry, among others. Within the sciences, we can distinguish at least the following: financial economics; market microstructure; behavioral finance; social studies of science; econophysics. While econophysics is located in mathematics and physics departments, financial economics, market microstructure, and behavioral finance are usually located in the finance departments of business schools. The social studies of finance is the less well institutionalized approach, usually located in sociology departments, and lacking the educational programs (such as master's, for instance) that the other approaches have.[1]

[1]Outside the sciences, we also encounter approaches claiming rigor and discipline. Technical analysis, for instance, enjoys relatively widespread usage among (professional and non-professional) traders, as well as some institutional structures, such as professional organizations and conferences. Technical analysis, however, is not regarded as a science and is not established as such within the academia. It does not claim a conceptual core—that is, a series of propositions and assumptions about human behavior and decision making, in the way the sciences of finance do.

These sciences of finance shouldn't be understood as mere discourses [39, p. 364], in the sense of providing (more or less realistic) accounts of how financial markets work. They include conceptual assumptions about what markets are, about the nature and character of participants, about the data to be processed, about the tools with which the data should be appropriately processed, as well as about the outcomes of data processing and the uses of such outcomes. Moreover, each of these approaches addresses a particular kind of audience: financial economics, market microstructure, and econophysics, for instance, address both academic and practitioner audiences. Behavioral finance and social studies of finance address mostly (different) academic audiences. To a growing extent though, behavioral finance has begun to find the interest of practitioners as well. Technical analysis is seen by academics as a non-academic discipline [39, p. 369] and, therefore, addresses almost exclusively audiences made of practitioners. (This, however, doesn't mean that its practitioners cannot be educated in a standard discipline such as financial economics.)

Historically speaking, the evolution of these approaches to finance has developed in partly parallel, and partly intersecting ways. Financial economics, the oldest of them, has emerged in the 1950s [14, p. 186]. Behavioral finance emerged in the 1980s, under the influences of decision-making theories developed in cognitive psychology [4, 45]. Market microstructure became prominent in the 1990s [36] as a consequence, among others, of technological changes in financial markets: the usage of computers in trading and the automation of transactions, which had taken off in the 1980s, drew attention to issues related to the structure of the order book and the execution of transactions. Econophysics emerged on a parallel path, in physics and mathematics departments, in the mid-1990s as well. Its practitioners publish mostly in physics journals [8, p. 993, 15].

What are then the objects of systematic inquiry for each of these perspectives, the methodological assumptions, data, outcomes, and prescriptions for these outcomes? How do these objects and assumptions impact assumptions about values, and how do they stand in relationship to each other?

2 Financial Economics

For financial economics, the dominant perspective is the efficient market hypothesis (EMH). While it has been regarded as an artificial construct (e.g., Howden [18, p. 8]), relevant in the context of this analysis is the object of systematic inquiry which emerges from this construct. This object is provided by the price changes of financial instruments (such as stocks), changes assumed as independent of previous periods and as conforming to a known probability distribution [18, p. 9]. Features such as the institutional conditions under which such price changes take place (e.g., the organization of exchanges) or the rules according to which orders are executed are irrelevant here. Minimal assumptions about the market actors (traders)—that they act according to expectations, or that they incorporate information in their decisions—are introduced as additional, secondary requirements, which remain largely unexplicated. Market actors are both a-psychological and a-social. They do not have to be concerned about diminishing attention, or about imitating each other, or about emotions. This is relevant with respect to what is at stake in this approach, namely the competition among investors to find the best investment prospects. Given the fact that there are few psychological or social elements impacting this competition, a state of equilibrium where arbitrage opportunities disappear will be quickly reached [5, p. 94]. The various versions of the EMH (weak, semi-strong, and strong) allow for different definitions of what counts as information (past prices, or publicly available information about companies, or everything), thus accepting or rejecting the usefulness of various analytical techniques such as technical, or fundamental analysis [5, p. 95]. Beyond that, though, the ultimate picture which emerges from this approach is that of a state of competition among market actors for finding investment opportunities.

This state of competition doesn't have to be justified—it is taken as a given, or as natural (see also Morgan [32, p. 579]. It doesn't have to be defined, or investigated closer—what does it actually mean to be in competition with others for investment opportunities? There are no explicit moral values at stake, or intervening in this competition. There is little concern, if any, about how real competitions take place, and how (stock) market competitions would compare with other types of competitions. The value of the EMH is that it provides both a summary picture of an assumed competitive state and a set of tools for a normative picture of decision making decisions within this competition, by stating that the search for anomalies in long term returns is illusionary [13, p. 304]. This means that, in the long run, participants in this competition cannot exploit consistently each other's weaknesses and that the competition itself will at least slow down, if not end, since all arbitrage opportunities will be exploited. (This also raises the question of the temporality of decision making in this competition—how long is long term?) In their search for best investment opportunities, market players do not have to be constrained by ethical considerations, since the choice is made purely based on information incorporated into prices. The value of financial economics itself is that it provides market actors with a toolbox which is adequate to their search.

3 Behavioral Finance

In contrast to mainstream financial economics, and as a reaction to it, behavioral finance (BF) emerged in the 1980s under the influence of prospect theory, a decision making approach developed in cognitive psychology (e.g., Tversky and Kahneman [45]). Behavioral finance does not assume market actors as being devoid of psychological features. Traders and investors have emotions, dwindling attention, and are influenced by each other in their decision-making. They can be overconfident, and they are influenced by illusions (e.g., Kahneman and Riepe [21], Della Vigna [12]). Decision making is influenced by diverse psychological factors (also called biases), which are understood as “imperfections” or departures from the assumptions of perfect attention, lack of emotions, of illusions, and the like. Since market transactions require decision making, it follows that the prices of financial securities will also be influenced by psychological factors, in addition to information. These psychological factors can be investigated in the laboratory or in field experiments. Their consequences can be observed in price patterns, giving rise to particular market phenomena or effects (also called anomalies by the proponents of the EMH).

While BF shares with mainstream financial economics an interest in analyzing price changes, it sees the latter as being patterned by the constantly recurring psychological “imperfections” of market actors [11]. Consequently, patterns of price variations become analyzable and interpretable based on such psychological features, a perspective which makes BF somewhat more sympathetic to technical analysis than the financial economics (the latter rejects technical analysis as completely non-scientific). While social interactions are acknowledged as influencing the decision making of market actors as well (e.g., Shiller [44]), the main focus is on the psychological features of market actors and the ways in which these impact decision making.

Behavioral finance has been accused of being a-theoretical, in the sense of not providing a general model of decision making in finance [13]. While prospect theory offers some general principles or notions of how human actors make decisions, these notions are not assembled in a deductive model. Rather, behavioral finance seeks to identify patterns of price movements, explain, and formulate prescriptions about how to exploit them (e.g., Jegadeesh and Titman [19], Grundy and Martin [16]).

The overall picture of financial markets we derive from behavioral finance is not substantially different from that of mainstream financial economics: markets are competitive situations in which players try to identify price patterns generated by the psychological features of market players and exploit them. While financial economics argues that arbitrage opportunities disappear in the long run, behavioral finance is not concerned with the this aspect, since the short term exploitation of price patterns is perfectly feasible. The value of a “science of finance” consists then in devising tools by which market players could exploit such patterns taking into account the psychological features discussed above.

Of course, one could raise here the following question: if market actors use tools of intervention based on identifying the price effects of psychological features such as emotions, or attention, or overconfidence, does the use of such tools make them less susceptible to be impacted in their decision making by inattention, or by overconfidence? In other words, is the use of such tools also a reflexive device that improves the psychological setup of market actors, by making them more aware of impact of inattention, or of overconfidence? If this were so, the value of behavioral finance would reside not only in formulating prescriptions about how to exploit specific price patterns but also, and perhaps more importantly, in changing the psychological makeup of market actors, by making them resemble more the ideal representation of rational actors provided by financial economics. Behavioral finance would have then a performative character [27], in the sense that its usage modifies the behavior of its practitioners, not by making them resemble more its predictions, but less.

Paradoxically then, while financial economics operates within the ideal of emotionless, fully attentive, perfectly calibrated market actors, it would be behavioral finance which brings real market actors closer to this approach. As I am not aware of any empirical research about the impact of behavioral finance on its users, this remains a question for future research.

Behavioral finance, though, has largely avoided questions related to how ethics impacts the decision making of traders. There are few, if any empirical studies on this aspect, as opposed to numerous studies about the impact of psychological features. We do not know whether altruism or reciprocity, for instance, impact trading decisions in any way, or whether ethical and psychological factors combine to any degree. To be fair, though, contract theory, which recognizes the significance of ethical choices for transactions, has recently begun to make inroads into behavioral finance (e.g., Aggarwal and Goodell [2]).

4 Market Microstructure

In contrast with behavioral finance, and also with mainstream financial economics, market microstructure analyses focus primarily on the following question: what happens to trading orders when they arrive on an exchange venue? What are the rules of their execution and how do these rules impact the behavior of market actors who send orders to the exchanges? The key assumption here is that market actors will act strategically and will use the execution rules of a particular exchange venue, as well as information about other orders being placed, as they arrive, to their advantage. They will seek opportunities for exploiting execution rules, as well as the information they have about other orders, without revealing though their intentions or all their information at their disposal [37, p. 260].

These strategic actions impact price and volume of transactions as they happen. The patterns of real time changes in price and volume can be analyzed and explained based on knowing execution rules, information about orders (i.e., price, volume, and characteristics such as limit vs. market), as well as on assumptions about the strategic behavior of market actors (e.g., Chang and Cheng [9], Ryu [42], Cerrato et al. [7]). While financial economics and behavioral finance are primarily concerned with the decision making of human actors, market microstructure acknowledges that market players do not necessarily have to be human. Market players can also be algorithms (i.e., software programs) designed to automatically place particular market orders or to identify particular price patterns (e.g., Guilbaud and Pham [17]). Algorithms can display strategic behavior too, since they are programmed by their builders to follow a particular set of prescriptions or principles that are not self-evident from the orders they place.

A significant difference then between the ways in which market microstructure, mainstream financial economics, and behavioral finance conceive of market players, is as follows: while the EMH operates within the confines of a benchmark model of perfect rationality in decision making, and behavioral finance within a set of assumptions about psychological "imperfections" (or "biases") which impact decision making, market microstructure takes actors to be opportunistic, strategic players, who will not (fully) reveal information or intentions, and will try to hide both.

Prices become then not so much an expression of the available information, or an outcome of psychologically imperfect decision making, as the expression of strategically revealed bits and pieces of information, used by players in advancing their game against opponents. Market microstructure is primarily interested in the order flow (that is, buy/sell orders as they arrive on the exchange venue, are queued up and executed according to the rules of the venue). Price patterns in the order flow can be explained on the basis of the players' strategic behavior, but are also a means of deciphering the latter. In contrast to financial economics and to behavioral finance, market microstructure is not so much interested in long term price variations, but in very short term ones. Types of data such as analysts' forecasts, or accounting data about specific companies (that is, fundamental data) are of less interest to market microstructure than data about real time (or tick by tick) price variations.

Since it conceives of market players as acting strategically, market microstructure does not engage with issues of ethics, or the extent to which ethical considerations might inform decision making. However, execution rules and regulations play a significant role, since they are constraints which flow into the decision making of traders. Moreover, market microstructure does not see market actors as limited to human beings—algorithms are included here as well. The use of algorithms in trading is not strange to ethical issues though, as debates have shown (e.g., McNamara [30]). The value of the approach resides in providing analytical tools for understanding the structure of the order flow, but also for interpreting the intentions of market actors as a premise for strategic decision making.

Financial economics and behavioral finance are largely technology-neutral, in the sense that, within their respective conceptual assumptions, technology does not make a difference with respect to the decision making of market players. Irrespective of the technology used by the latter for placing and executing orders—be it face to face in the trading pit, on the phone, or through computer software, decision making will remain medium-neutral, as it is either perfectly rational, or affected only by psychological "imperfections." For market microstructure, however, technology is significant (e.g., Mizrach and Neely [31]): the medium in which transactions are conducted impacts execution rules, the nature of market players, as well as the strategic decision making of participants. It is not the same if orders are placed in the pit or electronically: execution rules will be different on an electronic platform from those of the pit. Moreover, the strategic information players get is different in the pit from that of an electronic trading platform: in the pit, players orient themselves to glances, body postures, and shouts, while on an electronic platform they orient themselves to sequences of price and volume data, and to the rhythm in which this data arrives.

In fact, market microstructure itself as an academic approach has been very much facilitated by technology, and by the creation of databases including tick by tick prices. While attempts at microstructure analysis were already started in the 1980s, it is only in the 1990s, when stock exchanges made available their computer-based data, that the research took off [48, p. 142].

5 Social Studies of Finance

While all these three approaches have been institutionally located mostly in business schools and within financial economics, social studies of finance (SSF), the fourth under consideration here, appears to some extent to be an outlier. With a few exceptions, its proponents are not located in business schools, but in sociology and anthropology departments, and its assumptions are largely different from those of financial economics, behavioral finance, and market microstructure (although there are shared research interests as well).

SSF has emerged in the mid-1990s as a social science effort at understanding the significance of financial markets for society at large. While preoccupation with the stock exchange as a social institution or with financial markets was not new outside economics (e.g., Weber [47], Baker [3], Adler and Adler [1]), SFF added a series of new dimensions to this approach. First, SSF considers financial markets as social institutions and transactions as social activities. This has at least twofold implications: markets are not regarded in isolation from other social institutions, and price patterns are not regarded as the sole outcome of transactions. Relationships among market players, or group-endorsed views about the nature of trading as a social activities are also among the outcomes of transactions.

Second, SSF is preoccupied with the nature and types of knowledge at work in financial markets, and especially financial transactions, in relationship to both financial institutions and practices (e.g., MacKenzie [28], Muniesa [33]). This knowledge is neither reducible to price and volume information, nor assumed as identical with the benchmark model of perfect rationality, nor considered to be "tainted" by psychological imperfections. Financial knowledge includes tacit elements anchored in previous experiences. It has a practical rather than a theoretical character; it mixes formal and informal elements, and is shared within a group or community of practitioners. Groups, however, can develop variations on this knowledge, as well as proprietary forms, which they do not necessarily share with other groups. Therefore, SSF emphasizes rather the diversity of the forms of practical financial knowledge rather than assuming the existence of universal cognitive templates shared by all market players.

Third, SSF considers that the medium in which financial transactions are conducted has consequences not only with respect to the properties of said transactions (which are seen as social activities), but also with respect to financial institutions, and to the organization of the financial activities. Conducting transactions face to face in the trading pit or conducting transactions electronically is not the same. Interaction formats and technologies supporting the interaction of market actors have consequences with respect to price volatility, to the institutions supporting trading, and to the diffusion of financial instruments such as derivatives (e.g., Baker [3], Pardo Guerra [38], MacKenzie and Millo [26]). Therefore, SSF pays attention to the social forces and groups which are involved in developing trading technologies, and to the struggles and conflicts around these technologies (e.g., MacKenzie [29]). However, technology in financial markets is neither seen as having a linear trajectory, nor as following a path of relentless progress. Technology is seen as the site of struggles among market actors, and as impacting their behavior and decision making (e.g., Muniesa [33], Lepinay [24]).

Fourth, SSF considers that disciplines such as financial economics are not merely descriptions of an idealized state, or of a real one (in the case of behavioral finance, for instance). Social science disciplines dealing with financial markets are also modes of intervention in the market, in the following sense: the sciences of finance are preoccupied with constructing models (for instance, for identifying optimal investment opportunities), which will then be used by (at least some) practitioners. Repeated use of the models has the potential of changing trading practices, and hence the character of financial markets (e.g., Callon [6]).

Methodologically speaking, most SSF studies have been qualitative investigations of historical processes through which market institutions are transformed, and of social practices in contemporary financial institutions. While there is a rather limited number of qualitative studies in behavioral finance, financial economics and market microstructure studies operate exclusively with quantitative methodologies.

Similarly with market microstructure, SSF investigates how technologies impact the decision making of market players, and does not see them as consisting solely of human actors. While market microstructure focuses primarily on order execution and price variations, SSF pays attention to historical and institutional aspects (e.g., how specific technologies become established and how they change the character of market institutions, including here regulatory aspects) and to the social relationships among market players. In a way similar to behavioral finance, SSF sees market players as departing from the benchmark model of perfect rationality, but it is less interested in psychological "imperfections" than in how social relationships affect decision-making. This means that while behavioral finance and market microstructure have a more atomistic and individualistic view of market players, SSF sees them more as part of social networks and groups, and sees market activities as involving groups rather than isolated individuals.

While SSF does not always explicitly regard the choices of market actors as moral choices, there is an ethical component inherent to the approach. Since market actors are influenced in their decision-making by relationships, group activities, and reciprocal commitments or obligations (e.g., Knorr Cetina and Bruegger [22]), and since social relationships (by definition) imply moral constraints and obligations, trading decisions are not free of ethical implications. The latter are irreducible to following explicit normative prescriptions such as institutionalized codes of conduct in trading. The ethical implications of decision making consist rather in the reciprocal obligations which, while remaining mostly unspoken, have constraining character. These reciprocal obligations do not necessarily follow universalistic moral codes but are rather group-specific. They are also tied to group-specific valuation practices (e.g., MacKenzie and Spears [25]), which means that they have to be investigated in connection with the knowledge practices of various groups acting in markets. Thus, the ethical aspect of market behavior is not so much related to fiduciary elements (e.g., traders as optimizing profits on behalf of a principal), but to reciprocal commitments and obligations which constrain and shape market decisions (e.g., Knorr Cetina and Bruegger [22]).

Overall, then, SSF sees value not in functional terms (that is, not as being restricted to a particular function of trading, such as providing liquidity), but in terms of the practical actions of market actors, in terms of their knowledge practices, and in terms of the constraints and obligations that arise in relationship to these practices. At the same time, the value of SSF does not consist in providing a toolbox for practical actions in markets: SSF does not offer prescriptions for trading, or models based on which one can identify and exploit price discrepancies. The value of SSF is that it provides a broader perspective on financial markets as anchored in forms of institutionalized knowledge with larger cultural and social implications. Since it is difficult to imagine markets working without such institutionalized knowledge, and because the cultural and social implications of the latter are unavoidable, the value of finance is irreducible to functional elements (such as liquidity). The value of finance also consists in its connections to other forms of institutionalized knowledge (and hence to other social institutions), and in its consequences with respect to broader social issues such as justice, or inequality.

6 Econophysics

The fourth science of finance, econophysics, is also the only one that has not been developed within social science departments (including here economics departments and business schools). Institutionally speaking, econophysics is located within mathematics and physics departments, and thus formally separated from the social sciences (e.g., Gingras and Schinckus [15]. While the intellectual roots of econophysics are seen as laying in the work of Benoit Mandelbrot, among others [20, p. 326], its takeoff as a discipline happened in the mid-1990s, when the first conferences were organized, papers and edited volumes were published, followed later by establishing MSc and PhD educational programs, and by the creation of research institutes.

Institutionally speaking, econophysics has remained entirely separated from financial economics, and students of econophysics are not acculturated in the main tenets of financial economics, such as the EMH. As with financial economics, the development of econophysics has been made possible by the creation of large databases from financial trading and by the adoption of computing technologies [20, p. 323, 330]. The similarities between the institutionalization paths of financial economics and econophysics, respectively, also include the use of specific academic outlets for disseminating the own arguments (econophysics publishes mostly in physics journals), the downplaying of alternative theories as having less relevance, and the claim of greater empirical relevance of the own point of view [20, p. 334].

Methodologically, econophysics emphasizes Lévy distributions instead of the Gaussian distributions favored by the EMH. This difference is significant and has been seen as a point of contention. While Lévy distributions allow for infinite variance (the restricted version of these distributions doesn't though), Gaussian distributions do not allow for infinite variance. Within markets, infinite variance means that the prices of financial instruments can vary infinitely, which is more difficult to conceptualize. However, Lévy distributions also allow for a more refined analysis of extreme events and of anomalies, such as fat tails, volatility persistence, and volatility clustering [20, p. 337].

Data-wise, econophysics focuses on intraday and tick-by-tick price and volume data, sharing here an interest with market microstructure approaches. However, econophysics does not share the conceptual framework of financial economics, in the sense of not being interested in a set of assumptions about decision-making and the behavior of market players as the theoretical foundation of market analysis. Indeed, the separate institutional evolution of econophysics, with publication venues and educational programs distinct from those of financial economics, does not encourage the dialogue across disciplines [20, p. 344]. At the same time, as I have mentioned, students of econophysics do not get acculturated in the main tenets of EMH and therefore do not ascribe much significance to it.

Under these conditions, econophysics does not seem to emphasize the development of a theoretical framework of its own, concentrating instead on the empirical analysis of market phenomena. The disinterest in theory [20, p. 346] is, paradoxically perhaps, similar to the theoretical indifference manifested by technical analysis, which concentrates on the identification of price patterns.

Similarly to the financial economics, behavioral finance, and market microstructure, econophysics is not directly interested in broader issues about the value of finance, which should include ethical aspects. Neither does econophysics treat the value of finance primarily in functionalist terms. However, it sees its own value as in providing a more accurate toolbox for analyzing market anomalies such as volatility persistence or clustering, phenomena which cannot be easily explained under the methodological and conceptual assumptions of financial economics. In this respect, the value of econophysics is presented similarly to that of behavioral finance or market microstructure: a more accurate analytical approach to price variations, which does not resort to (unnecessary) theoretical assumptions.

7 The Value of the Sciences of Finance

What are then the ways in which sciences of finance conceive of their own value, and how do they understand their own boundaries, both in the sense of the own limitations, as well as in the sense of being distinct from each other? With respect to their internal distinctions, at least two sets of boundaries seem to be present: first, there is a division between approaches which operate with assumptions about individualistic human behavior, and approaches which do not resort to such assumptions. The theoretical cores of EMH, behavioral finance, and market microstructure include assumptions about individual decision-making and the factors shaping the latter. Irrespective of whether these factors are rational calculations of utility, psychological imperfections affecting such calculations, or game-like calculations, ultimately decision-making is predominantly individual. SSF and econophysics do not operate with assumptions about individual decision-making, though for different reasons. SSF's stance is that decision-making is rather influenced by institutional factors, as well as by group interactions, and therefore it is less individual than one might think. At the same time, for SSF decision-making is situational and not subjected to a universalistic model of rationality, or of deviations from rationality. Decision-making is bound by collectively developed knowledge practices that are related to material setups, or technologies generating particular types of information. For econophysics, however, decision-making is not seen as a necessary theoretical assumption. This does not mean that no decision-making takes place; it does. It just means that assumptions about decision-making are not necessary in order to analyze patterns of price variations. Price and volume are entities akin to physical particles; while market players may be behind them, we do not need to make particular assumptions about the background in order to analyze what is happening in the foreground.

This leads me to the second internal distinction, namely that between approaches which see markets as quasi-closed processes or entities, and approaches which do not. A quasi-closed process/entity in this case would be one that is self-sustained by its internal dynamics, which takes precedence over influences from the outside. Thus, for financial economics, behavioral finance, and econophysics, markets are quasi-closed processes: the internal dynamics takes precedence over any external influences, constraints, or links to other social institutions. For SSF and market microstructure, though, this is not necessarily the case. Market microstructure plays close attention to how external constraints (such as rules and regulations) influence decision making, while SSF analyzes markets as social institutions inextricably linked political and cultural arrangements, to education systems, or to technology.

The value of finance itself—that is, the value of the field of inquiry of all these disciplines—remains mostly implicit, in the sense that it is not systematically addressed. With the exception of SSF, this value is taken as self-evident. The mere existence of financial markets as a domain of activity justifies their investigation. Yet, a significant distinction is that between the utilitarian stance of financial economics (less so market microstructure, behavioral finance and econophysics) and the institutionalist stance of SSF. While financial economics sees the value of financial markets as consisting primarily in providing liquidity, for SSF the value of financial markets is not given by a single functionalist feature. Neither is the value of finance an invariant. The value of finance resides in how social and political institutions connect to markets, and in how these connections are made accountable for the society at large. Thus, the value of finance varies historically, depending on the links among institutions such as the state, firms, political parties, non-governmental organizations, the media, and markets. The value of finance depends, among others, on how institutions connect with and perceive financial markets; how markets are made accountable and legitimate with rapport to these institutions; how public perceptions of market activities are shaped.

Since it depends on inter-institutional links, the value of finance can be the locus of contestations and conflicts. It is not seen as the outcome of a historical deterministic and unilinear process, in which society increasingly acknowledges this value. While financial crises are addressed by other disciplines as well (e.g., Muradoglu [34], Reinhart and Rogoff [41]), they are a particularly relevant object of study for SSF. Crises are seen as revealing the struggles and contestations around the value of financial markets for the society at large, but also the struggles and contestations within markets themselves around notions of value (e.g., MacKenzie [28]).

How do these disciplines see the value of their own approach? Oftentimes, but not always, the value of the own approach is claimed in epistemic and methodological terms, namely as offering a more accurate perspective than that of competitors, or of using more adequate tools in the explanation of financial phenomena (e.g., Ray [40]). This makes approaches such as financial economics, behavioral finance, and econophysics contest each other's validity directly or indirectly, or claim that the own theoretical tenets haven't been disproven. For instance, proponents of financial economics have criticized both behavioral finance and econophysics for lack of a unitary theoretical framework, while financial economics in its turn has been criticized by the latter for operating with unrealistic assumptions or with a methodological toolbox that does not do justice to the realities of contemporary financial markets. One notable exception here is SSF, which does not claim to use more adequate tools, but to offer a more comprehensive perspective on finance.

To some extent, this situation is similar with that of scientific controversies from other disciplines, as documented in the sociology and history of science (e.g., Collins [10], Latour [23]). In ongoing controversies, epistemic authority and success depend on the parties' capabilities of mobilizing technologies of evidence (such as measuring devices) in favor of their standpoints, but also of accounting for "anomalies" in their data and for detecting such "anomalies" in their opponents' data. Up to now, at least, a significant amount of controversies have been taking places around accounting for anomalies (e.g., Fama [14]), less so around the mobilization of technologies for gathering evidence. We have seen less debates and contestations about what constitutes the "right" price data, for instance, than we have witnessed debates about how to account for market anomalies. With the rapid evolution of market technologies though, we should probably expect more and more debates around what constitutes the right kind of price data—for instance, tick-by-tick price data vs. quarterly or monthly price data—in relationship to the appropriate methodological assumptions for analyzing such data. Econophysicists, for instance, have made a point of using Cauchy and Lévy distributions as more appropriate, but have not insisted too much on the advantages provided by price recording technologies.

Overall, then, we can distinguish between internalistic and externalistic ways of claiming value for disciplinary approaches to finance: internalistic ways claim value based on methodological adequacy with respect to data, and on data accuracy (what kind of prices are the most accurate, and what kind of price recording is the most appropriate). The notion of data is understood here primarily, if not exclusively, as price and volume data, naturalistically produced by price recording technologies. Externalistic ways claim value based on the capacity to investigate connections between financial markets and other social institutions, connections which are not reducible to externalities (i.e., to the effects finance has on these institutions). Rather than being contradictory, the two value claims, internalistic and externalistic seem to be complementary. The complexity of financial transactions warrants internal investigations of the structure and dynamics of price variations. At the same time, this complexity has not only formal dimensions, but socio-cultural and institutional ones as well. Ignoring the latter would make much more difficult to justify the value of finance.

Overall, while tensions and contradictions exist among the sciences of finance, including here rivalries, it is difficult to argue that there is only one "right" social scientific approach to financial markets. The coexistence and parallel institutional evolution of finance disciplines can be seen not only as a sign of the importance of financial markets among modern social institutions, but also as an indicator that the complexity of markets cannot be dealt with by a single disciplinary approach. As this complexity does not give any signs of abatement, we should welcome diversity in the ways in which the social sciences investigate finance.

References

1. Adler, P., Adler, P. (eds.): The Social Dynamics of Financial Markets. JAI Press, Greenwich CT (1984)
2. Aggarwal, R., Goodell, J.W.: Culture, institutions, and financing choices. How and why are they related. Res. Int. Bus. Finan. **31**, 101–111 (2014)
3. Baker, W.: The social structure of a national securities market. Am. J. Sociol. **89**(4), 775–811 (1984)
4. Barberis, N.: Thirty years of prospect theory in economics. NBER working paper 18621 (2013)
5. Brague, G.: Neither efficient, nor animally spirited, but eventually adjusting. The stock market according to L.A. Hahn. Q. J. Aust. Econ. **15**(1), 89–119 (2012)
6. Callon, M. (ed.): The Laws of Markets. Blackwell, London (1998)
7. Cerrato, M., Kim, H., MacDonald, R.: Microstructure order flow. Statistical and economic evaluation of nonlinear forecasts. J. Int. Finan. Mark. Inst. Money **39**, 40–52 (2015)

8. Chakraborti, A., Toke, I.M., Patriarca, M., Abergel, F.: Econophysics review I. Empirical Facts. Quant. Finan. **11**(1), 991–1012 (2011)
9. Chang, Y.-C., Cheng, H.-W.: Information environment and investor behavior. J. Bank. Finan. **59**, 250–264 (2015)
10. Collins, H.: Gravity's Shadow. The Search for Gravitational Waves. University of Chicago Press, Chicago (2004)
11. DeBondt, W., Muradoglu, G., Shefrin, H., Staikouras, S.K.: Behavioural finance: Quo Vadis? J. Appl. Finan. **18/2**, 7–21 (2009)
12. Della Vigna, S.: Psychology and economics. Evidence from the field. J. Econ. Literat. **47/2**, 315–372 (2009)
13. Fama, E.: Market efficiency, long term returns, and behavioral finance. J. Finan. Econ. **49**, 283–306 (1998)
14. Findlay, M.C., Williams, E.E.: A fresh look at the efficient market hypothesis. how the intellectual history of finance encouraged a real 'Fraud-on-theMarket'. J. Post Keynesian Econ. **23**(2), 181–199 (2000)
15. Gingras, Y., Schinckus, C.: The institutionalization of econophysics in the shadow of physics. J. Hist. Econ. Thought **34**(1), 109–130 (2012)
16. Grundy, B.D., Spencer Martin, J.: Understanding the nature of risks and the sources of rewards to momentum investing. Rev. Finan. Stud. **14**, 29–78 (2001)
17. Guilbaud, F., Pham, H.: Optimal high frequency trading in a pro rata microstructure with predictive information. Math. Finan. **25**(3), 545–575 (2015)
18. Howden, D.: Fama's efficient market hypothesis and mises' evenly rotating economy. Comparative constructs. Q. J. Aust. Econ. **12**(2), 3–12 (2009)
19. Jegadeesh, N., Titman, S.: Profitability of momentum strategies: an evaluation of alternative explanations. J. Finan. **56**, 699–720 (2001)
20. Jovanovic, F., Schinckus, C.: Econophysics. A new challenge for financial economics? J. Hist. Econ. Thought **35**(3), 319–352 (2013)
21. Kahneman, D., Riepe, M.W.: Aspects of investor psychology. beliefs, preferences, and biases investment advisors should know about. J. Portfolio Manage. **24**(4), 52–65 (1998)
22. Knorr Cetina, K., Bruegger, U.: Global microstructures. The virtual societies of financial markets. Am. J. Sociol. **107/4**, 905–950 (2002)
23. Latour, B.: The Pasteurization of France. Harvard University Press, Cambridge MA (1988)
24. Lepinay, V.: Codes of Finance. Engineering Derivatives in a Global Bank. Princeton University Press, Princeton NJ (2011)
25. MacKenzie, D., Spears, T.: The formula that killed Wall Street': the Gaussian Copula and modelling practices in investment banking. Soc. Stud. Sci. **44**(3), 393–417 (2014)
26. MacKenzie, D., Millo, Y.: Constructing a market, performing a theory. The historical sociology of a financial derivatives exchange. Am. J. Sociol. **109**(1), 107–145 (2003)
27. MacKenzie, D., Muniesa, F., Siu, L.: Do Economists Make Markets? On the Performativity of Economics. Princeton University Press, Princeton (2007)
28. Mackenzie, D.: The credit crisis as a problem in the sociology of knowledge. Am. J. Sociol. **116**(6), 1778–1841 (2011)
29. MacKenzie, D.: Material markets: how economic agents are constructed. Theory Cult. Soc. **29** (2), 150–153 (2012)
30. McNamara, S.R.: The law and ethics of high-frequency trading. Minnesota J. Law Sci. Technol. (2015). SSRN: http://ssrn.com/abstract=2565707
31. Mizrach, B., Neely, C.: The transition to electronic trading in the secondary treasury market. Federal Reserve Bank of St. Louis working paper 2006–012 (2006)
32. Morgan, M.: Competing notions of 'Competition' in late nineteenth century American economics. Hist. Polit. Econ. **25**(4), 563–604 (1993)
33. Muniesa, F.: The Provoked Economy. Economic Reality and the Performative Turn. Routledge, London (2014)
34. Muradoglu, Y.G.: The UK crisis of 2008: What is real and what is behavioural? (2009). SSRN: http://ssrn.com/abstract=1373025 or http://dx.doi.org/10.2139/ssrn.1373025

35. Neate, R.: Queen finally finds out why no one saw the financial crisis coming. The Guardian (2012), 13 December, 2012
36. O'Hara, M.: Market Microstructure Theory. Blackwell, Malden MA (1995)
37. O'Hara, M.: High frequency market microstructure. J. Finan. Econ. **116**(2), 257–270 (2015)
38. Pardo Guerra, J.P.: Financial automation. past, present, and future, pp. 567–586. In: Knorr Cetina, K., Preda, A. (eds.): The Oxford Handbook of the Sociology of Finance. Oxford University Press, Oxford (2012)
39. Phoa, W., Focardi, S.M., Fabozzi, F.J.: How do conflicting theories about financial markets coexist? J. Post Keynesian Econ. **29**(3), 363–391 (2007)
40. Ray, R.: Econophysics. Finance, economics and physics. Appl. Econ. Lett. **18**, 273–277 (2011)
41. Reinhart, C., Rogoff, K.: This Time Is Different. Eight Centuries of Financial Folly. Princeton University Press, Princeton (2009)
42. Ryu, D.: The information content of trades. An analysis of Kospi 2000 Index Derivatives. J. Fut. Mark **35**(3), 201–221 (2015)
43. Sen, A.: Money and Value. On the Ethics and Economics of Finance. Edizioni Dell Elefante, Rome (1991)
44. Shiller, R.: Conversation, information and herd behavior. Am. Econ. Rev. **54**, 181–185 (1995)
45. Tversky, A., Kahneman, Daniel: The framing of decisions and the psychology of choice. Science **211**, 453–458 (1981)
46. Weber, M.: Economy and Society. University of California Press, Berkeley CA (1978)
47. Weber, M.: Stock and commodity exchanges. Theory Soc. **29/3**, 305–338 2000 [1896]
48. Wood, R.A.: Market microstructure research databases. History and projections. J. Bus. Econ. Stat **18**(2), 140–145 (2000)

Quantification Machines and Artificial Agents in Global Finance: Historical-Phenomenological Perspectives from Philosophy and Sociology of Technology and Money

Mark Coeckelbergh

Abstract This paper raises questions regarding the societal, cultural and ethical significance of finance, mathematics, and financial-mathematical technologies, discussing in particular the phenomenon of quantification as mediated by contemporary electronic information and communication technologies (ICTS). It first relates the history of mathematics to the history of financial technologies, and argues, inspired by Simmel and Marcuse, that from ancient times to now there seems to be an evolution towards increasing quantification not only in finance, accounting etc., but in modern society in general. It shows that scientific and technological changes have social and ethical consequences, as quantification creates more distance between people. The paper then analyzes and discusses current shifts of financial agency that exemplify what seems to be a moment of hyper-quantification through the use of ICTs: experiences of "the market" as an independent agent and money machines as artificial agents in high frequency trading—perhaps the only agents still able to cope with the data-loaded and hyper-quantified world we live in. Under these conditions it becomes more difficult to exercise responsibility. The paper concludes that while we must acknowledge the human character of finance and mathematics, there are real human and social consequences of quantification, in ancient times and today, for society and responsibility. It is therefore misleading to assume that financial technologies and mathematics are ethically neutral; more analysis of ethical and societal aspects is needed, also from an "outside" perspective.

Keywords Quantification · Ethics of finance · Sociology of finance · Philosophy of finance · Mathematics · Artificial agents · Phenomenology · Simmel · Marcuse · Responsibility · Distance · Philosophy of technology and media

M. Coeckelbergh (✉)
Department of Philosophy, University of Vienna,
Universitätsstrasse 7 (NIG), 1010 Vienna, Austria
e-mail: mark.coeckelbegh@univie.ac.at

E. Ippoliti and P. Chen (eds.), *Methods and Finance*,
Studies in Applied Philosophy, Epistemology and Rational Ethics 34,
DOI 10.1007/978-3-319-49872-0_10

1 Introduction

Quantification is widely used in mathematics and finance, as behavior of financial agents such as investors and markets is turned into numbers and modeled with the help of mathematics. Yet how this is done and its scope has changed over time, and the history of finance and mathematics is also a history of science and a history of technology, including a history of information and communication technologies (ICTs). Each of these "financial technologies" make possible a different phenomenology of markets, of financial transactions, and indeed of society. They also have consequences for how we do things.

This chapter does not contribute to financial modeling or mathematical theory, but approaches these fields from the "outside". It uses philosophy and sociology as well as a historical perspective in order to raise critical questions concerning the phenomenon of quantification and its ethical and societal significance. In particular, developing and drawing on recent work on the ethics of financial technologies [1] this chapter focuses on some of the philosophical and social problems connected to quantification through ICTs: what happens when our experience of, and action in, markets is mediated by information technology? How do these technologies change the phenomenology of markets and financial transactions? What are the moral and societal consequences of quantification and of these technologies?

The latter are important questions given the societal consequences of finance, and hence also of financial and mathematical theory and technologies. It is assumed in this paper that people working in these fields want to do their work in a morally and socially responsible way. Therefore, it is vital to make explicit (potentially) darker sides and consequences of these practices and phenomena. This can contribute towards more critical reflection on finance and society, and may aid efforts towards more responsible practice and innovation in the sector.

The first section of the paper offers a brief history of financial technologies and shows that these technologies have always had ethical and societal consequences. In particular, using and expanding Simmel it is argued that there seems to be a tendency towards increasing quantification in finance and in society, that this also means increasing distance between people and between people and products, and that this evolution is crucially related to the history of financial technologies. In addition Marcuse is used to issue a warning about how quantification may change the lifeworld and contribute to the quantification and bureaucratization of society.

The second section turns to contemporary global finance and in particular its electronic ICTs. It is argued that, phenomenologically speaking, the new technologies and quantification tend to create increasingly autonomous non-human financial agents: "the market" appears to us as a being in itself, and financial algorithms effectively constitute artificial agents which trade on our behalf—something which seems to be needed once finance, aided by ICTs, takes on a global shape and reaches such high speeds. It is argued that this creates problems for exercising responsibility, given serious limitations to human control and knowledge. Like all automation technology, contemporary financial technologies and the mathematics and science that go with it are in danger of creating more epistemic and social distance.

It is concluded that while this process of increased quantification and distancing does not necessarily mean a dehumanization of finance, it is clear that we have to deal with the reality of these new technologies and agents, and that—to use a metaphor of calculation which has shaped our thinking—there might be a significant moral and societal cost to them.

2 A Brief History of Quantification and Financial Technologies and an Argument About Distance Based on Simmel and Marcuse

2.1 A Brief History of Quantification and Financial Technologies

From ancient times to today, finance, mathematics, and financial technologies and artefacts have always been co-developing. New financial-economic realities required new science and technology, but also vice versa: new mathematics and new techniques made possible new forms of finance. While I have no space here to do justice to the details of this fascinating history, let me indicate some significant financial-technological developments.

From the time of the agricultural revolution it was possible to have 'stock', since there was more than people needed for immediate consumption. This had all kinds of social consequences, such as most likely increased competition, inequality, and domination by (male) elites. It also had an implication for technological development: stock needed to be managed, counted, calculated. At the same time new forms of social organization developed such as cities and empires, together with bureaucracies and taxation. Stock can be distributed, collected, centralized, taxed, etc. New technologies were needed to do this. An important financial technology was and is writing. In Mesopotamia, Egypt, China, and other ancient centres of civilization people needed to write down quantities for bookkeeping and administration. Writing systems were invited, for instance Phoenician and Greek writing systems, and of course also *numbers*. This made possible ancient versions of accounting and what Martin calls management information systems: techniques for quantifying stocks and flows of goods, combined with measuring of time [2, p. 43]. The abacus, for instance, is a well-known ancient calculating tool: a tablet or counting frame that was already in use in ancient Mesopotamia and Egypt.

Another important financial technology was and is of course money. Some connect the birth of money to the need for a medium of exchange. Here the idea is that first there was barter, but this only works as long as both parties want each other's goods. If this so-called 'double coincidence of wants' does not happen, money provides a handy medium of exchange. Morgan has argued that to deal with this problem a standard means of exchange was found, which first could take the form of valuable goods such as cattle, cloth, or cereals [3, p. 11] and later also silver, lead, copper, bronze, and so on. Some means of exchange, such as cowry shells, were used across continents. Other authors emphasize the role of debt as the origin of money (see for example [4–6]). Here there was an evolution from personal obligation towards other persons to more formal forms of debt, associated with the activities of what we now call "banks." Already in ancient Babylon temples took deposits and gave loans, in 11th century China people already used paper money, and in Renaissance Italy and Flanders bank notes were used by foreign exchange dealers, so-called *banchieri* who dealt on benches [7]—the origin of our word "bank". Goldsmiths also played a role in the history of credit and banking, as they issued receipts for gold they kept.

2.2 An Argument About Distance Based on Simmel and Marcuse

In these histories of financial and mathematical technologies, we can discern not only technological and scientific changes, but also social changes. In particular, put in spatial terms the development of these technologies and the related quantification process seem to go together with increased distance: from families and small communities to large bureaucracies and empires, and from personal, informal social relations to impersonal, formal social relations. As I have argued in *Money Machines* [8], from a phenomenological-geographical point of view this financial-mathematical development can be conceptualized as a process of *distancing*: distance between people and goods, but also increasing distance between people. To explain this, let me use the work of Simmel.

In this philosophy of money [9], Simmel argued that money, as it mediates the exchange relation, functions as a kind of bridge (it relates people, even strangers, when they engage in exchange) but at the same time the technology or medium also creates distance: between people and between people and objects In order to fulfill its function, Simmel argued, money has to be impersonal, detached from specific content and value. This renders the social relation impersonal and also alienates us from the value of goods. This distancing is directly related to quantification. In the course of history every qualitative difference becomes quantified, and money becomes dematerialized: first it has the form of valuable objects, then it becomes coin money, paper money, and so on.

This process of quantification and dematerialization reaches its peak in modernity, when money becomes the symbol of 'the modern emphasis on the quantitative moment': objects are 'valued only to the extent to which they cost money' and money quantifies since it is 'free from any quality', cut off from the relevant relationships. [9, p. 279] Once quantified, the object is no longer of interest. According to Simmel, this makes people indifferent towards objects, but also towards one another. Money, as the symbol of abstraction and quantification, turns people more calculating. In our lives are engaged in various processes of quantification. Simmel writes:

> The money economy enforces the necessity of continuous mathematical operations in our daily transactions. The lives of many people are absorbed by such evaluating, weighing, calculating and reducing of qualitative values to quantitative ones. [9, p. 444]

The money economy promotes this kind of relations with goods, but also with people. Simmel compares the nature of money with prostitution: through money we objectify each another, degrade each another to 'mere means' [9, p. 377].

In Marcuse we also find criticism of quantification. In *One Dimensional Man* [10] he argues that the quantification of nature separated science from ethics, the world of rationality and the world of values. The latter, he says, are seen as not real because they cannot be quantified or scientifically described. It is not part of the objective world, it is not real. For ethics this means that 'outside this rationality, one lives in a world of values, and values separated out from the objective reality become subjective' [10, p. 151]. In other words, Marcuse suggests that ethics is seen as not real and as subjective; what "counts" is, literally, what can be counted, what can be quantified. But, like Simmel (and later also Habermas), Marcuse suggests that in reality this objective world is not separate at all but changes the lifeworld. In the end not only nature becomes transformed into a 'technical reality' [10, p. 158]; society and human beings also become rationalized and quantified, and then controlled and manipulated by means of technology. Quantification makes this control and manipulation possible. Quantification is a form of abstraction, and this abstraction has consequences for how we perceive the world and for our practices.

Marcuse specifically mentions mathematics. Commenting on Husserl, Marcuse argues that 'mathematization' [10, p. 141], required for technological-rational thinking, hides the 'pre-scientific basis of science in the world of practice (*Lebenswelt*)'. It creates the illusion of a free-standing, autonomous, symbolic truth: 'an absolute ideational reality, freed from the incalculable uncertainties and particularities of the *Lebenswelt*' [10, p. 167]. But in reality, Marcuse says, 'it remained a specific method and technique for the *Lebenswelt*. Based on Husserl, he writes: 'The ideational veil (*Ideenkleid*) of mathematical science is thus a veil of symbols which represents and at the same time masks (*vertritt* and *verkleidet*) the world of practice' [10, p. 166]. Mathematics makes possible a 'specific concrete experience of the *Lebenswelt*—a specific mode of "seeing" the world. This then leads to particular kind of practical relation to the world: the domination of nature, but also domination in society. First, social reality is seen in an objective and calculable way, which misses out its 'mysterious and uncontrollable character' [10, p. 172]. Second, here too mathematics leads not only to a different perception but also to different practice. Again technology plays a role: 'in the medium of technology, man and nature become fungible objects of organization' [10, p. 172]. Quantification makes possible domination and control by means of numbers. The result is that 'The world tends to become the stuff of total administration, which absorbs even the administrators' [10, p. 172].

Simmel's and Marcuse's conclusions may well be exaggerated and far too pessimistic, but they invite us to further critically reflect on the nature of quantification and of the financial technologies, and attend us to their potential social and moral implications. In the next section I explore what this approach means for contemporary financial technologies and contemporary quantification.

3 ICTs and Artificial Agents in Finance: Markets, Algorithms, and the Question Concerning Responsibility

3.1 Markets and Algorithms in Global Finance and Their Social and Ethical Consequences

Today we still live in a money economy, but one that is globalized and involves a range of new technologies. In particular, electronic information and communication technologies (ICTs) make possible global markets, where similar processes of quantification and hence distancing are at work. The numbers on the screens of traders express values and relations, but they abstract and impersonalize them. The concrete people and relations disappear, are invisible. Moreover, they create a world of numbers (the world of mathematics, science, economics, and finance), which hides the *Lebenswelt* and sets up the world of values (ethics) as separate from finance. It thus misleadingly suggests that its mathematical-financial operations are ethically neutral and have nothing to do with people's experience, which is seen as "subjective".

For instance, traders at a stock exchange do no longer perceive an immediate link to the goods they trade and the people they (literally) deal with. They are part of a "technical" world, which seems unconnected to the lifeworld. Like money, electronic trading platforms thus function as media that connect people and people with goods (the bridge function). Yet at the same time, as quantification machines, they also act as screens which create distance. This distance is ethically problematic, since the social consequences of finance remain hidden. In addition, the traders themselves are controlled and manipulated by the system.

Moreover, in so-called high-frequency trading and related practices, trade is delegated to algorithms. The financial world becomes a world of 'quants': mathematics experts who program computers to analyze and trade on the market. In that world of numbers, it seems, machines are more at home than humans. Technology is used to create what Simmel called 'pure quantity in numerical form' [9, p. 150]. According to Simmel, this is the nature of money. Today it is thus 'technically feasible to accomplish what is conceptually correct' [9, p. 165]: money becomes pure quantity, without a material basis. Electronic money and cryptocurrencies such as Bitcoin generated by computer algorithms are pure symbols. The financial-economic world is turned into a world of pure quantities, detached from place. The global world of electronic money and media is in this sense a "utopia": a non-place. Everything becomes quantity, information, data. Humans can no longer cope with the speed of trade, only machines can still act at this summit of quantification.

In addition, financial products become so complex that only mathematicians know what is going on and, perhaps very soon, only computers can understand them and deal with them. The financial city has become an island and now drifts off: it alienates itself not only from all the people elsewhere on the globe and indeed even creates distance in the same city who are affected by the trade but lack any control over it; it also alienates itself from human traders. Finance then becomes a gigantic machine, served by goods and people. Its technical rationality makes possible global domination, in which both traders and stakeholders are trapped.

Even in cases when trade is (still) done by humans, it is mediated by electronic trading platforms. This means that direct contact between traders on the market or on the trading floor is replaced by numbers on a screen. Whereas for instance on exchanges trading in the 'pit' involved 'full-body experiences' including hand signaling and shouting [11, p. 263], this is replaced by abstract, de-personalized and disembodied data. This also changes the nature of the social interaction and the sense-making. Trade now appears to be about numbers only, not about people.

Furthermore, financial markets, which are supposed to be anchored in real transactions and real human beings, appear to us as separate, non-human entities: new bests and monsters, new artificial agents which are abstracted from goods and people, and reign in the world of money. "The market" thus creates new forms of global domination. Again technologies—including numbers, computers, and screens—play an important role here. As Knorr Cetina remarks about technologies on trading floors, the market becomes the market-on-screen which takes on 'a presence in its own right':

> From the traders' perspective, and from the perspective of the observer of traders' lifeworld, the dominant element in the installation of trading floors in globally interconnected financial institutions is not the electronic infrastructural connections … but the computer screens … The market on screen takes on a presence and profile in its own right … It is not simply a 'medium' for the transmission of pre-reflexive interactions' [12, p. 129]

Technologies are not mere instruments; they also shape our perception and our world. In this case, mathematics and financial technologies shape a world which appears to be ruled by "the market" as it appears on our screens: in the trading rooms but also on TV and elsewhere. Again this may be a hyperbolic claim. Finance remains human, in the sense that it is still humans who program the computers and make part of the decisions. In this sense to say that it is a nonhuman world is an overstatement. But overstatements can alert us to real processes, in this case processes of abstraction and quantification, which create what seems a "symbolic" world (divorced from material and social reality); but this world and its numbers have real social and moral consequences.

This analysis is not only applicable to finance. Increasing quantification processes means that *everywhere* in society, at least in so far as it is a modern society, increasing numbers of people are involved in the work of quantification. Quantification and bureaucratization happen not only in finance but in all sectors, including health care, education, and the private lives of people. But in all modern social institutions, increasing quantification and abstraction means that social relationships tend to become less personal and calculating. In so far as financial technologies and mathematical techniques function as quantification machines and money machines, they may contribute to these social and moral developments.

3.2 Implications for the Exercise and Ascription of Responsibility

Consider again the creation of abstract markets and the delegation of trade to machines: what are the consequences for responsibility?

It seems that it becomes rather difficult to exercise and ascribe responsibility. Who is responsible for the actions of "the market", an artificial agent which seems to steer humans rather than the other way around? And who is responsible for the actions of trade algorithms? As Aristotle already knew, there are two conditions for responsibility: control and knowledge. We know that acting responsibly is only possible if we have some control and if we know what is going on. But with increased automation and increased quantification, can we still fulfil these conditions and exercise responsibility? If we conceptualize knowledge in purely quantitative terms, as in big data science, and delegate gaining knowledge to machines such as search engines and big data miners, then what kind of knowledge do we humans have, and is it enough for responsible action in the world of finance and beyond? For instance, do citizens and politicians have enough knowledge of the world of finance in order to make good democratic decisions concerning its practices? Even if we do not embrace everything Simmel and Marcuse say about the social and ethical consequences of quantification, at the very least these questions need to be asked.

Numbers, money, and calculating techniques are great inventions. They can do a lot of fantastic things for us. But it is worth considering what else they do. They do not only have a purely "technical" function. They also shape our (life)world, our thinking, and our social relations and social institutions. The technology may become the end and, as McLuhan said, the medium becomes the message. Money, numbers, accounting, and other financial media and techniques are not neutral; they crucially shape what it is to be human today.

4 Conclusion

Humans have always invented new technologies and have developed mathematics and finance to help them with various kinds of activities (e.g. trade and exchange) and with trying to reach various goals. But science and technology have always done more than acting as a docile servant. They have also changed our practices and re-shaped our aims. They have made possible new activities and new cultures. They have contributed to the creation of entirely new worlds and civilizations. This chapter has argued that this is also true for financial technologies and mathematical techniques, today and in the past: they have always been bound up with social change and they have changed the economy, finance, and society. Drawing on Simmel and Marcuse, it has been shown what kind of changes we may consider here: it has been argued that current financial-mathematical developments contribute to more quantification and that this may have social and moral consequences: more impersonal relations to goods and people, a misleading conception of science and finance as ethically neutral, an alienated understanding of markets, more bureaucracy and administration, and conditions that undermine the possibility of exercising and ascribing responsibility.

I conclude that if we want a better understanding and a more comprehensive evaluation of the social and ethical significance and potential implications of finance and mathematics, it is advisable to not only consider "inside" perspectives of these sciences and practices and the many advantages they have brought us, but also critically study the relations between finance and mathematics, technology, ethics, and society. A philosophical and sociological angle such as the one presented here may, "from the outside", contribute to this aim. But the outside is also an inside: this is about what people experience and do, inside and outside of finance and mathematics.

References

1. Coeckelbergh, M.: Money Machines: Electronic Financial Technologies, Distancing, and Responsibility in Global Finance. Ashgate, Farnham/Burlington (2015)
2. Martin, F.: Money: The Unauthorized Biography. The Bodley Head, London (2013)
3. Morgan, E.V.: A History of Money. Penguin, Harmondsworth (1965)
4. Hudson, M.: The archeology of money: debt versus barter theories of money's origins. In: Wray, L.R. (ed.) Credit and State Theories of Money. Edward Elgar Publishing, Cheltenham (2004)
5. Hart, K.: Notes towards an anthropology of money. Kritikos **2** (2005)
6. Graeber, D.: Debt: The First 500 Years. Melville House, Brooklyn (2011)
7. Ferguson, N.: The Ascent of Money: A Financial history of the World. Allen Lane/Penguin Books, New York (2008)
8. Coeckelbergh, M.: Money Machines: Electronic Financial Technologies, Distancing, and Responsibility in Global Finance. Ashgate, Farnham/Burlington, VT (2015)
9. Simmel, G.: The Philosophy of Money, 3rd edn. ed. Frisby, D., trans. Bottomore, T., Frisby, D. Routledge, London ([1907] 2004)

10. Marcuse, H.: One-Dimensional Man: Studies in the Ideology of Advanced Industrial Society. Routledge, London ([1964]1991/2002)
11. Zaloom, C.: Ambiguous numbers: trading technologies and interpretation in financial markets. Am. Ethnologist **30**(2), 258–272 (2003)
12. Knorr Cetina, K.: From pipes to scopes: the flow architecture of financial markets. In: Barry, A., Slater, D. (eds.) The Technological Economy. Routledge, London (2005)

Dark Data. Some Methodological Issues in Finance

Emiliano Ippoliti

Abstract The nature of the data of financial systems raises several theoretical and methodological issues, which not only impact finance, but have also philosophical and methodological implications, viz. on the very notion of data. In this paper I will examine several features of financial data, especially stock markets data: these features pose serious challenges to the interpretation and employment of stock markets data, weakening the 'myth of data'. In particular I will focus on two issues: (1) the way data are produced and shared, and (2) the way data are processed. The first raises an internal issue, while the second an external one. I will argue that the process of construction and employment of the stock markets data exemplifies how data are theoretical objects and that 'raw data' do not exist. Data are not light and ready-to-use objects, but have to be handled conceptually and technically very carefully and they are a kind of 'dark matter'. *Dark data*, for the note.

1 Introduction

The nature and the role of data in the financial systems, especially in stock markets, raise several theoretical and methodological issues. These issues, on one hand, affect the very theoretical status of finance itself, while on the other reignite a philosophical debate, that is the one about the notion of data and their 'ladenness'[1]—i.e. the fact that they are conceptual products, the end-point of a theoretical construction and not 'neutral' starting-point of it (see also [14]). This is one of the reasons why the study of financial systems is interesting also from a philosophical viewpoint. And, in turn, why a philosophical reflection can be useful for finance.

[1]See in particular [3, 4, 27, 29].

E. Ippoliti (✉)
Department of Philosophy, Sapienza University of Rome,
Via Carlo Fea 2, Rome, Italy
e-mail: emi.ippoliti@gmail.com

E. Ippoliti and P. Chen (eds.), *Methods and Finance*,
Studies in Applied Philosophy, Epistemology and Rational Ethics 34,
DOI 10.1007/978-3-319-49872-0_11

In effect, the process of construction of data in stock markets exemplifies how data are theoretical objects in nature and that 'raw data' do not exist. Data are not light, stable, ready-to-use objects, but a kind of 'dark matter'. *Dark data*, for the note. They are the outcome of specific interpretations of financial phenomena and dynamics, and require a lot of work in order to be put in use, if any, and in order to read inside them. Data are not a form of zero-degree kind of knowledge, light and neutral. This is a very important point, since finance is one of the realms of the so-called 'Big Data' approach, whit its naïve revival of the inductivist idea (see [2, 22]) about data and the construction of knowledge—namely that we do not need conceptualizations and theories when we have so much information.

In this paper I will focus on two issues: (1) the way data are produced and shared (Sect. 2), and (2) the way data are processed by agents and their dark spots and areas (Sect. 3).

As concerns the first issue, I set out to show how data are endpoints and not starting points of the working process of markets. I will examine how data emerge from models and theories and I will discuss their *performativity* and what this might imply (Sect. 2.1), how the specific nature of these data requires different methods (Sect. 2.2), and how data are expression of underlying structures that characterize them in a way that pose serious limitation, or at least caveats, to the employment and exploitation of certain dataset (Sect. 2.3).

As concerns the second issue, I will examine how, once data are produced and shared, their processing is not so simple, since the *internal* features of the agent who process them affect both the way they are read and the conclusion that can be drawn from them (Sect. 3.1), and I will face the problem concerning *what* data to use in order to draw conclusions about financial phenomena (Sect. 3.2).

2 Ways of Constructing Data

Financial data, especially the ones coming from stock markets, are *constructions* that embed a lot of technology, processes, and models (). When we see financial data on a display, we are looking at the outcome of mechanisms built by employing specific rules, sequences and conceptual models. This is one of the main reasons, for example, why the study of the *microstructures* of markets has become more and more important. I will examine three *internal* issues of these constructions.

2.1 Models Generating Data

Saying that data are the outcome of models, theories and mechanism means at least two things. First, it means that a piece of data is the outcome of rules and mechanisms that produce it and they are not 'neutral' with respect to it, i.e. their outcomes. As a matter of fact these rules and mechanism affect the data in several ways. Since the data become what the financial agents have to be act on, the understanding of rules and mechanisms of their composition and aggregation is essential. A stock example comes from the study of *markets microstructures*.

Second, the use of certain models and theories might *perform* the markets and, accordingly, generate certain markets dynamics and resulting data. This idea is at the core of a markets feature labeled as *markets performativity* (and *counter-performativity*), which draws on the concept of *reflexivity* (self-fulfilling and self-cancelling prophecies, see [8]).

The basic idea behind performativity is a conceptualization of markets as a set of practices—or better a set of practitioners that employ certain tools (see [5, 18, 20]). These tools are technical and conceptual in nature (theories, models or technologies) and might affect, and shape, the markets and their behavior. The 'affection', when happens, is very specific, in the sense that the put in use of these tools (and the relative decision-making) modifies the behavior of its practitioners, the reality of markets, in a way that makes the markets resemble the theory. More in detail, such a modification can be broken down into two kinds: *performativity*, when the use of these tools makes the markets resemble the descriptions given by them, and *counter-performativity*, when we have the opposite outcome. So the main tenet of performativity theory is that our ideas (models and theories) about markets shape their dynamics, their *reality*. Quoting Callon, the "discourse contributes to the construction of the reality that it describes" ([6], p. 316), the discourse do precede the reality.

Even if performativity, and the use of term 'perform' is criticized (see e.g. [21]), it opens a interesting issue. Let us discuss it using as an example the *behavioral finance*. The rise of this approach, the increasing adoption of this conceptualization of markets by practitioners, outlines a performative scenario that might affect the data.

Behavioral Finance (BF) tells us that market actors, i.e. traders and investors, take decisions under the influence of several psychological factors and processes (biases and heuristics). Some of these processes are considered as imperfections, that is breaks of the perfect rationality assumption. Lacks of attention, emotions, biases, influence the decision-makers and markets dynamics, and accordingly their data are the outcome of these decisions. Now the performative question is the following: the employment of behavioral finance conceptualization in trading activity is reflexive or not?

If the practitioners, markets actors, employ BF, it means that they are trying to spot the effects on prices of these psychological factors. But in doing this they would take decision in a way that makes them more aware of these factors. If it so, the use of BF would change, and improve, their psychological approach and settlement, by making them less and less similar to description given just by BF, that is making them more rational, less affected by emotions, by lack of attention, overconfidence, fear, and so on (see the several biases, heuristics, and fast and frugal processes in H&B and F&F traditions as Sect. 3). In other words, BF would have a counter-performative effect. Thus, the markets dynamics, and the resulting data, would reflect the change produced by the use of the BF tools. Bottom line: "paradoxically then, while financial economics operates within the ideal of emotionless, fully attentive, perfectly calibrated market actors, it would be behavioral finance which brings real market actors closer to this approach" ([28], p. 157). In this case the adoption of a specific conceptualization of markets, i.e. models and theories of BF, would generate a specific set of data, which would be not the starting point, but the end point of models and theories that would be embedded in the data, literally.

2.2 *Dark Spots*

Most of the financial data available to us come in a very specific form. They have an important pieces of information missing—a kind of *dark spot*: who trades what. In effect most of the orderbooks we can access are without the trading-account identifiers (see e.g. Fig. 1).

This is just a macroscopic example of a dark spot in the data, the one having a major theoretical impact. In effect stock markets are plenty of these missing pieces of data—dark spots. Another example is the number of orders visible in the orderbook of a specific exchange venue. A further example is the use by the so-called informed traders of *odd lots* to hide trades from the markets (see [26]). Even if the SEC recently required odd lots reporting to the tape, "a variety of other data, such as Rule 602 (trade execution quality) statistics, still do not include odd lots, and historical data remain incomplete. Such missing data are a natural concern to researchers" ([25], p. 267). This fact opens a series of issues. First of all it makes it very difficult, if not impossible, to recognize sequences of actions by the same trader or firm (either human or algorithmic). Secondly, and even more importantly, this makes the financial systems a very peculiar and interesting field to investigate. On one side, this feature of the data, the *dark spot*, seem to display a theoretical weaknesses. On the other side, it is a challenge. In effect, it has been argued both ways. In the first case, the absence of such a pieces of information is considered critical, since it seems to prevent a deep understanding of financial phenomena and dynamics. Without a grain fine understanding of these systems it is impossible to produce findings, predictions and descriptions about it. In the second case, the absence of this piece of data is considered not so critical, since the financial systems con be understood also with this pieces of the puzzle missing (see also [12, 13]).

Moreover it is worth noting that, as noted by Donald MacKenzie, we have an additional problem: "researchers employed by regulatory bodies, who do have

Time	Quantity	Bid	Ask	Quantity	Time
10:23:14	100	23.34	23.36	250	10:03:17
10:21:12	19	23.33	23.36	1000	10:04:01
10:24:01	7000	23.3			

Fig. 1 An example of data in an order book

access to account identifiers, find the task of unraveling patterns [...] computationally, and perhaps conceptually, close to intractable" ([19], p. 20). Of course it does not mean that the problem is not treatable and solvable, but it points out the possibility that this is not a critical piece of information for cracking financial systems (especially their algorithmic part) and shows that we can gain knowledge about financial systems also without this piece of information—or that we can't gain knowledge also with it.

This very line of thought shapes one influent way of accounting for this issue, that is the one put forward by Econophysics (see e.g. [32, 37]). In effect, Econophysics adopts an emergentist approach that, in part, seems to solve the problem: you do not need this piece of information (the identifier, who trades what) to account for certain properties of the financial systems. Or better, you can still understand certain properties of the financial systems, and according to this approach also the most relevant one, without this specific data. These properties are the ones that shape the 'collective' behavior and so the understanding of an individual, finer, level is not important. The description and forecast of the collective behavior of course emerge from the aggregation of the individual level behavior. But this does not imply that if we improve the understanding of the individual level, so will our understanding of the collective level.

A core idea of econophysics approach is the analogy between the financial systems and earthquakes—and seismology in general (see in particular [7, 33, 32][2]). Earthquakes are singularities, i.e. critical points, preceded by foreshocks (and followed by aftershocks see Fig. 2), and in the same way, financial singularities, like crashes, would be critical points preceded by characteristic foreshocks.

In both cases, since we cannot have access to and directly measure a variable (stressing forces in seismology, individual transactions in finance), we have to rely on a theoretical account that measures and draws inferences indirectly, statistically, and collectively.

An interesting point of this approach is the fact that it tries to offer a reading key of certain markets dynamics, in particular crashes and flash crashes, and it also tries to provide a mathematical model for the identification of critical events, namely a model based on log-periodic functions, for the note (see [37]). Bur it also presents several limits in data-fitting.

In effect, the main thesis of this model is that in times of a speculative bubble, a financial index increases as a power law decorated with log–periodic oscillations and ends up with a crash—the climax of the so called Log-Periodic Power Law (LPPL) signature. Sornette discusses few historical examples of financial singularities that fit it, like the 1929 crash (see Fig. 3).

But there are other data that seem to undermine this account. Let us look for example at the S&P 500 index data in Fig. 4.

[2]It is no coincidence that Sornette was a geophysicist before entering the study of finance and stock markets.

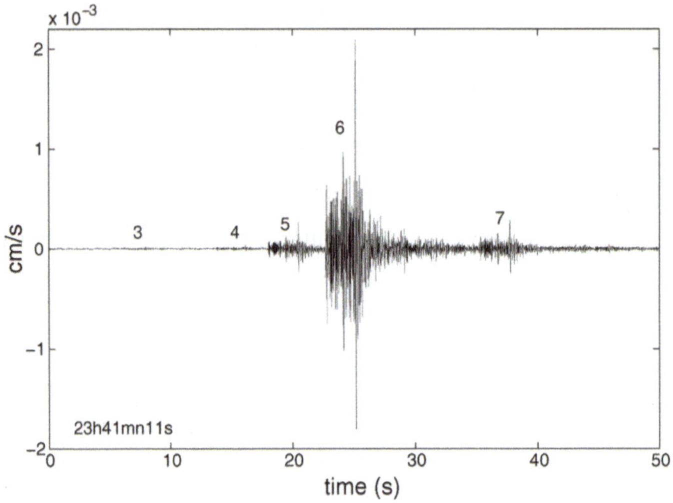

Fig. 2 Foreshocks and aftershocks of a critical event

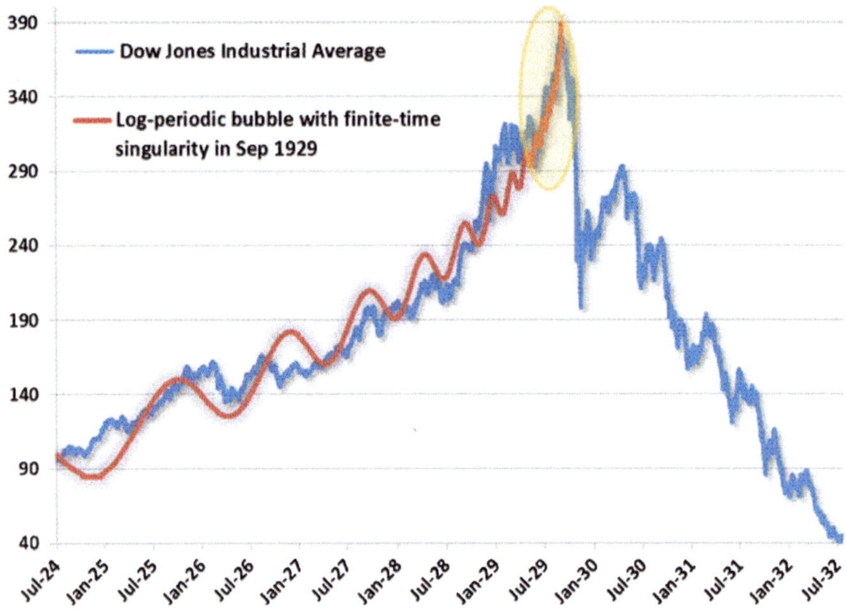

Fig. 3 The 1929 crash

These data project a financial singularity for April 2013. Unfortunately, no singularity occurred in April 2013, as you can see in Fig. 5.

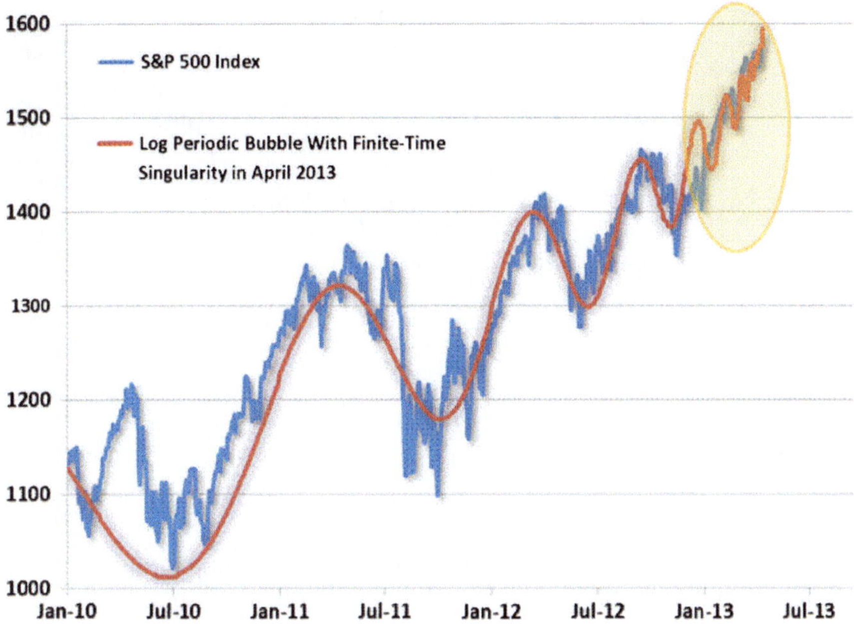

Fig. 4 S&P 500 index data Jan-2010/April 2103

This shows how can be difficult to make sense of data presenting this type of dark spots.

2.3 Structures Generating Data

Markets structures, that is the way markets are designed both at macro and micro level, shape the data in a several, and sometimes also subtle, fashion. These underlying structures change over time, might remain hidden in part or completely to a large part of the markets actors, and affect both the production of the data and the way to process and use them (by comparing them, aggregating, etc. see also Sect. 3.2). These structures, of course, are the production of an idea of the markets and its functioning, that is a conceptualization of it. Accordingly such a conceptualization is embedded in the data produced by the systems implemented and designed on it.

At a macro level, the structure of financial markets has changed a lot in the last few decades, and the resulting data reflect these different configurations of it. In the 1960s, for instance, the markets were regulated systems with controlled and limited cross market transactions. Basically a set of national financial systems were connected by few operators buying and selling transnationally, that is across national frontiers and across the exchanges and by a few national assets markets. Moreover,

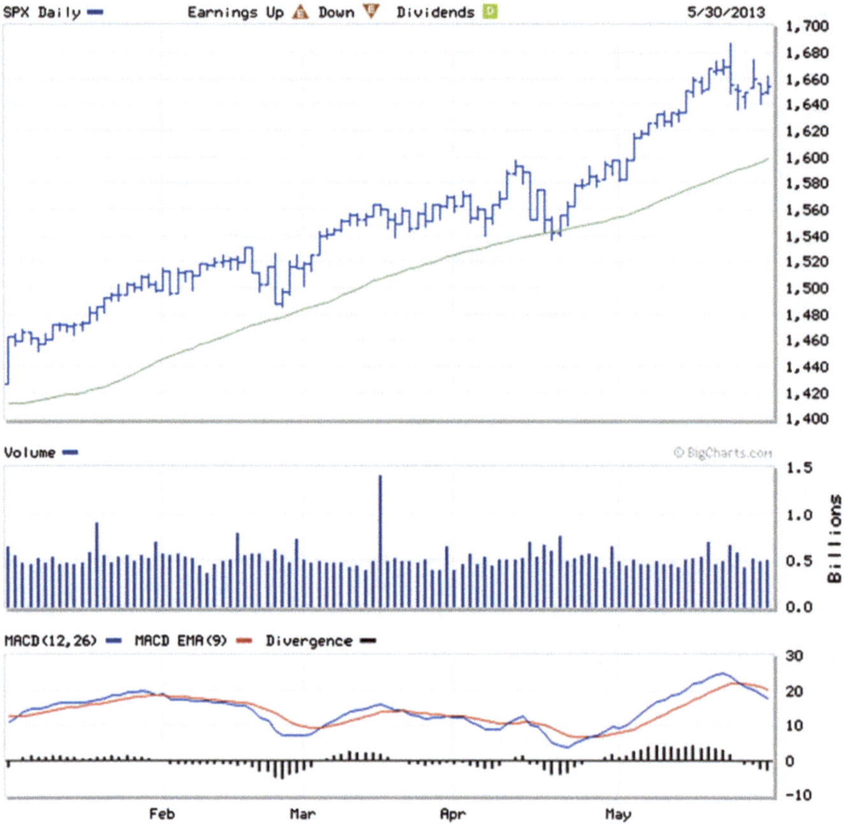

Fig. 5 S&P 500 index data after April 2103

the data for prices and volumes were *not* collected in real time, and were *not* low latency. Progressively a new design emerged, whereby, as noted by Strange [34], we now have a structure in the form of global system where the several national markets are separated only by physical distance, but work as a whole. This global financial system is now both larger and more influential than the various national financial systems. "The balance, in short, has shifted from being a predominantly state-based system with some transnational links to being a predominantly global system with some local differences" ([34], p. 260.)

Additionally in the very last few years a new design has emerged: a market composed of multiple trading venues connected by means of rules over access and trade priority. Especially in U.S. and E.U. this fragmentation has grown a lot. Moreover the data for prices and volumes nowadays are collected in real time, and are low latency. Bottom line "the current market structure is thus highly competitive, highly fragmented, and very fast. It is also dominated by the trading of high frequency traders, who by some estimates make up half or more of all of trading volume. Understanding what high frequency traders do is crucial for comprehending why markets today are so very different from times past" ([25], p. 258.)

Of course we can still see similar or even the very same 'patterns' emerging from dataset resulting from these different structures, but since they are the outputs of internal *machineries* that can be also very different, their use and comparison is controversial, if not useless or even misleading. When we say that the dynamics of prices of, e.g., 80s and 90s are similar ore even the same (and on this basis we produce certain predictions on futures trens in prices) we have to be very aware of their structural differences. Since data express historical processes, i.e. markets structures, they have to be continually and carefully contextualized, and you can treat different dataset as homogenous only in specific circumstances.

After all, past data are the result of policies, regulations, rules, etc. at work at that specific period. If you wish to use past data in order to understand what is going on today, it is not enough to know what, just to give an example, current policy makers will do. You also need to know the features of monetary policies at work beneath past data. This fact of course might undermine the idea of using a mathematical-statistical analysis as a way of understanding the markets in depth. Moreover, this prevents us from using a given dataset as a tool to understand and make forecasts about other datasets that expresses different structures, even if they look very similar. They are heterogeneous objects, or better expression of very heterogeneous structures, and they cannot tell a lot about each other.

At a micro level, on the other side, the structure of financial markets has changed a lot in the last few years, and the resulting data reflect these different configurations of it. One of the main engines of this transformation is the technological turn of financial systems (basically, algorithmic trading and HFT), and its effects on the rules of trading and order execution, the is the *market microstructure* (the study of the process and outcomes of exchanging assets under a specific set of rules). More and more, the data resulting after this technological turn embed specific conceptualizations, strategies and designs of financial markets (see [24] for the discussion of an ontological issue in a technical settings). For instance, as noted by O'Hara [25]:

> High frequency trading is strategic because it maximizes against market design. Exchange matching engines become the focal point of high frequency strategies, meaning that how the market is structured becomes very important. HF strategies can be quite complex, but so, too, now are the strategies that other traders elect, in part because they need to optimize in a market that contains HF players. And the exchanges as well act strategically, opting for new pricing models and market designs to attract (and in some cases deter) particular volume to their trading venues. As a result, trading is now different, and the data that emerges from the trading process is consequently altered ([25], p. 257.)

A remarkable example of the role and impact of the microstructures on financial dynamics and data is the Flash Crash occurred 5 May 2010, and more in general all the recent flash events in stock markets. Alteration of data is not the only possible outcome here. As a matter of fact, recent analyses of these controversial events (see for instance Aldrich et al. [1]) show how there are multiple levrecent flash events in stockels and steps needed to produce a piece of data (e.g. the price of a stock), which can interact in a way that can even generate uncertainty about the data. Or better, a data could be an unstable, unsure object, the endpoint of a process that can go wrong and give erroneous or unaligned (in time and space) outcomes. It follows that the rules and mechanisms by which data are a constructed, aggregated and disclosed to the market agents in effect is crucial. Since these data, in turn, are the building blocks of inferences and decision of traders and investors, and then produce actions and reactions, they shape the behavior of markets. The data are literally different because of the effect of the rules of executions and the technology employed by traders, investors and exchange venues.

3 Ways of Processing Data

In the previous paragraphs I have examined some issues arising from internal features of financial data, that is issues *inside* the construction of the data. In the following paragraphs I will examine issues that are external to the data—i.e. *outside* the process of construction of the data. These issues arise *after* the data are produced and built. More precisely, once data are produced and shared, their processing and use is not so simple and neutral for at least two reasons. First, because of the features of the agents processing and reading them, which affect the way they are read and, accordingly, change the conclusions that can be drawn from them (Sect. 3.1). Second, because the choice about *what* data to use in order to understand financial phenomena can be very tricky and noisy (Sect. 3.2). Therefore, also this external process displays dark areas, which make the financial data a very interesting issue from both a methodological and a theoretical viewpoint.

3.1 Dark Processing of Data

Once the data are produced and disclosed to the public, they are of course processed and employed in order to draw conclusions and take decisions. This data-processing can be done by humans, machines or a combination of them. Of course machines implement strategies designed by humans, so they cannot be totally different from the ones followed by humans. One of the main findings of evolutionary psychology, in particular two of its main approaches such as the H&B[3] and F&F[4] traditions, is that also the interpretation of data can be darkened, affected, by the features of the agent processing them.

[3]See Tversky and Kahneman [35, 36], Kahneman [15], Kahneman and Tversky [16].

[4]See Gigerenzer and Todd [9], Gigerenzer et al. [10], Gigerenzer and Selten [11].

Broadly, the H&B tradition argues that the task of data-processing by humans is darkened by emotions and psychological features, like loss aversion or risk aversion. The influence of these factors generally produce heuristics and biases leading to mistakes and errors, that is violations of optimization process. An optimization process, in this case, is the one defined by the Rational Choice Theory. The F&F school, in addition, points out that these *darkening* interferences arise from both the quantity and the form of data, which might conflict with the 'natural' (i.e. as made up by evolution) ways by which agents process them. Therefore the way by which data are presented to the subject is crucial for a correct or an incorrect elaboration of them.

More in detail the H&B approach displays how, and to what extent, the use of heuristics and biases interferes with the selection of the option that would maximize the classics expected value of conventional rational choice theory. The H&B school maintains that heuristics and biases work in such a way to produce decisions and inferences, and accordingly prices of financial securities, that contradict optimal rationality as formalized by the Expected Value Theory (EVT). The principal ways of affecting and darkening the inferential and decisional process recognized are heuristics like 'availability', 'representativeness', and 'anchoring and adjustment' (see [35, 36]). In effect the H&B tradition gives many examples of how, under certain circumstances, those heuristics and biases produce a miscomputation of probabilities, leading to the selection of the wrong consequences or possible end states. The consequences of the use of these heuristics and biases are reflected in prices patterns: they can trigger particular phenomena, like crashes and bubbles, which are seen as anomalies or failures by other approaches. In this way the H&B school displays how the cognitive features of actors prevent them to use in a proper way the information and data that they already have and trust, or take for granted.

Of course, the H&B school recognizes also the fruitful role of heuristics: the use of heuristics does not necessary generate mistakes and, contrariwise, it produces mostly good inferences or decisions. But under specific circumstance they trigger misleading inferential and decisional processes. This happens when the right, or better conclusions do not follow from the most readily processed cues, since these heuristics tend to ignore just the harder-to-process factors—those that can generally be ignored with little cost in order to produce accurate decision, but that sometimes turn out to be crucial.

The F&F school, as noted before, maintains that these problems derives mostly from the specific amount and form of the data—the way of presenting them. On this precise issue, the H&B and F&F schools tend to disagree, as the F&F tradition argues that by using a different quantity or form, the processes leading to mistakes would not be triggered.

As a matter of fact, first of all the F&F school argues that too much information do not improve the traders ability to make good decisions because such an overload of information will be expensive, too expensive, to process, in terms of both time and cognitive resources. Bottom line: all these data are distractors and generate poorer decisions. They misguide the attempt of a trader to locate the best piece of information while applying natural and effective heuristics like "take the best" or "follow the crowd". So a reduction of the amount of information put into the data disclosed to agents is the best way of triggering natural and effective heuristics, the ones leading to the best, or better, choice. In the specific case of financial systems, the F&F school argues that financial systems are capable of producing the useful cues most of the times, but heavy or improper regulations, amassing data, will darken them and then will push traders and investors to ignore even *all* the disclosed information in some cases. So an excessive amount of data disclosed to markets participants will push them to take decision and draw conclusions as if they were completely ignorant. Or, in some case, this circumstance will trigger compensatory decisions that lead to counterproductive effects.

Secondly, the F&F school argues that the way by which data are presented to agents will affect and alter their inferential or decisional process. Of course F&F school examines human agents. A stock example is the assessment of risk: in this case the use of data in the form of frequencies instead of probabilistic ratios (see Gigerenzer and Todd [9]) does not trigger poor decisions. Contrariwise, it is only when we employ the specific form of probabilities and percentages that we can incur in typical mistakes like the overestimation of risks.[5] Of course is not the case that inferences or decisions made on the basis of a better form (frequencies) are always necessarily superior to those based on a less 'natural' form (probabilities or percentages): the adoption of the 'better' form of data could not produce effect at all.

3.2 A 'Dark Matter': The Identification of Useful Data

A major concern with the employment of financial data comes from the identification of the kind of data and variable to use in order to understand financial phenomena. A stock example is the explanation and forecast of a financial recession. The attempt to find indicators, precursors and data foretelling recessions is a long-standing problem, which is now approached also with the help of the big data—and the huge amount of correlations that they allow to find. The noise in these data is so strong that it makes very difficult to spot leading indicators, that is a group of variables capable of signaling certain events in advance. The possible indicators that seems to fit past data are so many that the underdetermination of hypotheses by data grows exponentially.[6] It is really a *dark* matter.

Roughly speaking, here we face two interconnected theoretical issues.

The first can be defined as the one of the 'positional variables', the fact that reliable variables and indicators change from one economic or financial cycle to another one. Their value and reliability, if any, is positional, is time-sensitive. Big Data are more and more used to spot correlations the might identify causal

[5]This type of mistake derives from the fact the agents tend to neglect the base-rate information.
[6]See e.g. Quine [30], Newton-smith [23], Laudan [17] on this point.

indicators, but this task is very difficult. In effect, even when we find variables that appear to be as leading indicators in one economic cycle, they turn out to be as lagging indicators[7] in the next cycle. As noted by Silver, just to give an example, "of the seven so-called leading indicators in a 2003 Inc. magazine article, all of which had been good predictors of the 1990 and 2001 recessions, only two—housing prices and temporary hiring—led the recession that began in 2007 to any appreciable degree. Others, like commercial lending, did not begin to turn downward until a year after the recession began" ([31], p. 179). This holds also for the famous Leading Economic Index (LEI), a composite of ten economic indicators published by the Conference Board, which works intermittently.[8]

Moreover, since most of time these leading indicators are a set of variables (they are not jut one or two), the problem of their internal relation arises. As a matter of fact, since the global economic and financial systems continually evolve, the relationships between different economic variables can change over time. Thus a positional issue arises both externally, that is in the relation between variables and the financial cycles, and internally, that is in the relation between variables and other variables of the same indicators.

The second issue is a reflexive one. Since the indicators are, in turn, data that are employed as a base to draws conclusions and take decision (i.e. implement policies), they can change after they are measured and targeted, making them less reliable. More in detail, when a given policy is applied and begins to act on a particular variable, this variable might loose its value as a reliable economic indicator. A simple example is the couple house prices–economic health. If a government promotes actions that have the effect of raising house price, they might well increase, but they will no longer be good measures of overall economic health.

This fact shows how a seemingly unquestioned assumption, namely the notion that there are dependent and independent variables, well-fixed inputs and outputs that can be clearly separated from one another, turns out to be controversial in economics and in financial systems in particular. Of course this makes difficult to shed light on reliable ways of using data.

4 Final Remarks: Dark Data, Exploitation and Methods

These features of the data constructed and used in the financial systems have theoretical as well as practical implications.

---

[7]A lagging indicator is a variable that changes only after the economy follows a particular pattern or trend, or only after it experiences a large shift. Lagging indicators can confirm long-term trends, but they do not predict them. Stock examples are the unemployment rate, corporate profits and labor cost per unit of output.

[8]LEI has declined two months before recessions, as well as in other occasions where the trend was the opposite. For instance in 1984, it went down for three straight months, meaning a recession, while the economy continued go up.

As concerns the latter, the presence of this *darkness* in the data is of great practical concerns. The dark areas, in effect, are potentially exploitable and profitable. They provide competitive advantages and, in some case, they also might ease markets manipulations.

In order to illustrate the exploitable features providing competitive advantages and manipulations, I will discuss the example of markets fragmentation, and its reflection in the data. The fact that nowadays we have a markets structure without a central market (it is characterized by several venues), has changed, and is still changing, the way of trading and accordingly the resulting data. Such a fragmentation generates several sophistications into the trading environment. First, it generates new dynamics, since traders have to search for liquidity across many venues and markets, and the ability to do so at high speed provides a valuable competitive advantage.

Moreover, such a fragmentation generates arbitrage opportunities across markets—since prices might not always be the same in the several venues. Even more, it favors high speed trading in a subtle fashion. Since the several exchanges venues provide, by paying, direct feeds of their trading data, the traders operating at higher speed can see the market before, and with more clarity, than others (e.g. traders relying on standard consolidated tape data). Moreover, there is a further issue, a liquidity issue: since high speed traders can submit and cancel orders faster than most of the rest of the markets, it is difficult to spot liquidity across such fragmented markets. One of the main consequence of these new features, emerging from fragmentation and amplified by algorithmic trading, is that buys and sells are not the main data to look at in order to understand the markets dynamics. Time patterns, additions and cancellations to the orderbook, trade sequences, volumes, are examples of data on which strategies are built and inferences drawn.

But above all, this fact has interesting theoretical implications, namely that in order to make sense of price behavior, we have to look at data at across markets, and not just within individual markets. Orderbooks are interconnected, and so are price behavior and order flows. In some case the data coming from single markets could tell us very little, if any. They could give us only noise. But the right stream of data across different markets could provide us real patterns and reveal underlying structures and dynamics. The nature of market data is changing just as a result of such inter-relations between markets. The understanding of these inter-markets processes is difficult from a theoretical viewpoint and requires new approach, methods and empirical bases.

Eventually, the whole cognitive enterprises aiming at understanding financial systems strongly undermines the idea of the so-called "myth of data" (see Sellars [29]). This idea shapes a quite common tenet about the data and the acquisition of knowledge in general. This tenet broadly states that all we have do in order to gain new knowledge is to collect data and employ a set of appropriate procedures (a stock examples is some kind of generalizations) to get a hypothesis, which is the base for the construction of a theory that explains the regularities found in the data. This is the belief, often implicit, that the data are light, and the information is clear. In other words they provide us with a zero-degree of knowledge: they are neutral, independent, and objective.

Unfortunately, this idea is very difficult to sustain both theoretically and practically. As we have shown, especially in financial systems, the data are always the result of a theoretical operation, that is an interpretation of reality, and then the outcome of a conceptual construction and its specific purposes, which is also practical in nature. What we call 'data' is the result of the way by which we cut out a portion of 'reality' according to specific aims, so of a conceptual move. In the specific case of finance, data turns out to be not only the result of cutting out 'reality', but also a way of shaping 'reality' by cutting it out.

References

1. Aldrich, E.M., Grundfest, J., Laughlin, G.: The Flash Crash: A New Deconstruction (2016). SSRN-id2721922
2. Bacon, F.: Novum Organon. London (1620)
3. Basu, P.K.: Theory-ladenness of evidence: a case study from history of chemistry. Stud. Hist. Phil. Sci. **34**, 351–368 (2003)
4. Bogen, J., Woodward, J.: Saving the phenomena. Philos. Rev. **97**, 303–352 (1988)
5. Callon, M.: Introduction: the embeddedness of economic markets in economics. In: Callon, M. (ed.) The Laws of the Markets. Blackwell, Oxford (1998)
6. Callon, M.: What does it mean to say that economics is performative? In: MacKenzie, D., Muniesa, F., Siu, L. (eds.) Do Economists Make Markets? On the Performativity of Economics, pp. 310–357. Princeton University Press, Princeton (2007)
7. Feigenbaum, J.A.: A statistical analyses of log-periodic precursors to financial crashes. Quant. Finan. **1**(5), 527–532 (2001)
8. Flanagan, O.J.: Psychology, progress, and the problem of reflexivity: a study in the epistemological foundations of psychology. J. Hist. Behav. Sci. **17**, 375–386 (1981)
9. Gigerenzer, G., Todd, P.M.: Simple Heuristics That Make us Smart. Oxford University Press, New York (1999)
10. Gigerenzer, G., Hertwig, R., Pachur, T.: Heuristics: The Foundation of Adaptive Behavior. Oxford University Press, New York (2011)
11. Gigerenzer, G., Selten, R. (eds.): Bounded Rationality: The Adaptive Toolbox. Dahlem Workshop Reports. The MIT Press, Cambridge, MA (2001)
12. Ippoliti, E.: Methods and Finance. A View from Inside. In: Ippoliti, E., Cheng, P. (eds.) Methods and Finance. A unifying view. Springer (2016)
13. Ippoliti, E.: Methods and Finance. A view from Outside. In: Ippoliti, E., Cheng, P. (eds.) Springer (2016)
14. Ippoliti, E., Cellucci, C.: Logica. Egea, Milano (2016)
15. Kahneman, D.: Thinking, Fast and Slow. Macmillan, London (2011)
16. Kahneman, D., Tversky, A.: Prospect theory: an analysis of decision under risk. Econometrica **47**(2), 263–291 (1979)
17. Laudan, L.: Demystifying underdetermination. Minnesota Stud. Philos. Sci. **14**, 267–297 (1990)
18. McKenzie, D.: An engine, not a camera. How Financial Models Shape Markets. MIT Press, Cambridge (MA) (2006)
19. McKenzie, D.: How Algorithms Interact: Goffman's 'Interaction Order' in Automated Trading, draft (2016)

20. MacKenzie, D., Muniesa, F., Siu, L. (eds.): Do Economists Make Markets?. Princeton University Press, On the Performativity of Economics, Princeton (2007)
21. Maki, U.: Performativity: Saving Austin from MacKenzie. In: Karakostas, V., Dieks, D. (eds), EPSA11 Perspectives and Foundational Problems in Philosophy of Science, pp. 443–453. Springer, Dordrecht (2013)
22. Mill, J.S.: A System of Logic Ratiocinative and Inductive. Longman, London (1843)
23. Newton-Smith, W.: The underdetermination of theory by data. Aristotelian Soc. Suppl. **52**, 71–91 (1978)
24. O'Hara, M.: What is a quote? J. Trading **5**(2), 10–16 (2010)
25. O'Hara, M.: High frequency market microstructure. J. Financ. Econ. **116**(2), 257–270 (2016)
26. O'Hara, M., Yao, C., Ye, M.: What's not there: odd lots and market data. J. Finan. **69**(5), 2199–2236 (2014)
27. Popper, K.: Objective Knowledge. Oxford University Press, London (1972)
28. Preda, A.: In: Ippoliti, E., Cheng, P. (eds.) Springer (2016)
29. Sellars, W.: Empiricism and the philosophy of mind. In: Feigl, H., Scriven, M. (eds.) Minnesota Studies in the Philosophy of Science, vol. 1, pp. 253–329. University of Minnesota Press, Minneapolis (1956)
30. Quine, W.V.: On empirically equivalent systems of the World. Erkenntnis **9**, 313–328 (1975)
31. Silver, N.: The Signal and the Noise. Penguin Books, New York (2012)
32. Sornette, D.: Why Stock Markets Crash. Princeton University Press, Princeton (2003)
33. Sornette, D., Bouchaud, A., Bouchaud, J.P.: Stock market crashes, pre- cursors and replicas. J. Phys. I France **6**, 167–175 (1996)
34. Strange, S.: Finance, information and power. Rev. Int. Stud. **16**(3), 259–274 (1990)
35. Tversky, A., Kahneman, D.: Rational choice and the framing of decisions. J. Bus. **59**(4), S251–S278 (1986). Part 2: The Behavioral Foundations of Economic Theory
36. Tversky, A., Kahneman, D.: Advances in prospect theory: cumulative representation of uncertainty. J. Risk Uncertain. **5**, 297–323 (1992)
37. Zhou, W.Z., Sornette, D.: Evidence of a Worldwide stock market log- periodic anti-bubble since mid-2000. Phys. A, **330**(3), 543–583(41) (2003)

FSC
www.fsc.org
MIX
Papier aus verantwortungsvollen Quellen
Paper from responsible sources
FSC® C105338

If you have any concerns about our products,
you can contact us on
ProductSafety@springernature.com

In case Publisher is established outside the EU,
the EU authorized representative is:
Springer Nature Customer Service Center GmbH
Europaplatz 3, 69115 Heidelberg, Germany

Printed by Libri Plureos GmbH
in Hamburg, Germany